FOOD PROTECTION AND SAFETY

FOOD PROTECTION AND SAFETY

LYMAN F. THOMASSON
EDITOR

Nova Science Publishers, Inc.
New York

For permission to use material from this book please contact us:
Telephone 631-231-7269; Fax 631-231-8175
Web Site: http://www.novapublishers.com

LIBRARY OF CONGRESS CATALOGING-IN-PUBLICATION DATA

Available upon request.

ISBN: 978-1-60692-187-6

Published by Nova Science Publishers, Inc. ✝ New York

CONTENTS

PREFACE

Societal and demographic changes, the globalization of the food supply, new threats and communication issues require new approachs to food protection and safety. The chapters in this new book present important new information in this area which is crucial to everyone.

"Americans enjoy unprecedented choice and convenience in filling the cupboard today, but we also face new challenges to ensuring that our food is safe. This Food Protection Plan will implement a strategy of prevention, intervention and response to build safety into every step of the food supply chain."

The Food and Drug Administration (FDA) is responsible for ensuring the safety of roughly 80 percent of the U.S. food supply, including $417 billion worth of domestic food and $49 billion in imported food annually. The recent outbreaks of E. coli in spinach, Salmonella in peanut butter, and contamination in pet food highlight the risks posed by the accidental contamination of FDA-regulated food products. Changing demographics and consumption patterns underscore the urgency for effective food safety oversight. In response to these challenges, in November 2007, FDA and others released plans that discuss the oversight of food safety. FDA's Food Protection Plan sets a framework for food safety oversight. In addition, FDA's Science Board released FDA Science and Mission at Risk, which concluded that FDA does not have the capacity to ensure the safety of the nation's food supply.

The Food and Drug Administration (FDA) is responsible for ensuring the safety of roughly 80 percent of the U.S. food supply, including $417 billion worth of domestic food and $49 billion in imported food annually. Changing demographics and consumption patterns along with an increase in imports have presented challenges to FDA. At the same time, recent outbreaks, such as E. coli from spinach and Salmonella from tomatoes, have undermined consumer confidence in the safety of the food supply. In November 2007, FDA released its Food Protection Plan, which articulates a framework for improving food safety oversight. In January 2008, GAO expressed concerns about FDA's capacity to implement the Food Protection Plan and noted that more specific information about the strategies and resources needed to implement the plan would facilitate congressional oversight.

Numerous federal, state, and local agencies share responsibilities for regulating the safety of the U.S. food supply, which many experts say is among the safest in the world. Nevertheless, critics view this system as lacking the organization and resources to adequately combat foodborne illness, which sickens an estimated 76 million people and kills an estimated 5,000 each year in this country. The 1 10th Congress may face calls for a review of federal food safety agencies and authorities, and proposals for reorganizing them have been introduced (e.g., H.R. 1148; S. 654). Other bills that would alter some agency responsibilities include S. 1274, H.R. 2108, S. 1082, H.R. 1600, H.R. 2629, and S. 887. Among the issues likely to arise are whether reform can improve oversight, and the cost to industry, consumers, and taxpayers. This chapter provides a brief introduction to the system and the debate on whether reorganization is needed.

Foodborne illness continues to present a public health challenge in the United States (U.S.). The Centers for Disease Control and Prevention (CDC) estimates that, in the U.S. each year, 76 million people become sick, more than 325,000 people are hospitalized, and 5,000 people die from foodborne illness. The estimated economic cost of foodborne illness in terms of pain and suffering, reduced productivity, and medical expenses is substantial, in the range of $ 10-83 billion each year. Foodborne illness is associated with a range of foods, including fresh produce.

In November 2007, the Food and Drug Administration released the Food Protection Plan (FPP), an integrated strategy for protecting the nation's food supply. The FPP addresses both food safety and food defense for domestic and imported products. FDA is working collaboratively across the agency to address the three core elements of protection: prevention, intervention and response. This summary provides an update on the accomplishments under the FPP in the first six-months and an overview of activities planned to begin now or in the near future.

In November 2007, the Food and Drug Administration released the Food Protection Plan (FPP), an integrated strategy for protecting the nation's food supply. The FPP addresses both food safety and food defense for domestic and imported products. FDA is working collaboratively across the agency to address the three core elements of protection: prevention, intervention and response. This summary provides an update on the accomplishments under the FPP in the first six-months and an overview of activities planned to begin now or in the near future.

Societal and demographic changes, the globalization of the food supply, new threats and communication issues require a new approach to food protection. In May 2007, U.S. Department of Health and Human Services Secretary Michael O. Leavitt and Andrew C. von Eschenbach, M. D., Commissioner of Food and Drugs, charged the FDA with developing a comprehensive Food Protection Plan to address both unintentional and deliberate contamination of the nation's food supply.

The Food and Drug Administration (FDA) will begin a Tomato Safety Initiative in the Summer of 2007. The Initiative is a collaborative effort between FDA and the state health and agriculture departments in Virginia and Florida, in cooperation with several universities and members of the produce industry.

The U.S. Food and Drug Administration (FDA) will begin a multi-year Tomato Safety Initiative to reduce the incidence of tomato-related foodborne illness in the United States.

In: Food Protection and Safety
Editor: Lyman F. Thomasson

ISBN 978-1-60692-187-6

Chapter 1

FOOD PROTECTION PLAN AN INTEGRATED STRATEGY FOR PROTECTING THE NATION'S FOOD SUPPLY

"Americans enjoy unprecedented choice and convenience in filling the cupboard today, but we also face new challenges to ensuring that our food is safe. This Food Protection Plan will implement a strategy of prevention, intervention and response to build safety into every step of the food supply chain."

Michael O. Leavitt

A MESSAGE FROM THE COMMISSIONER

As a physician and the Commissioner of Food and Drugs, protecting America's food supply is extremely important to me.

American consumers have one of the safest food supplies in the world, but the world is changing and we know it can be safer. New food sources, advances in production and distribution methods, and the growing volume of imports due to consumer demand call for a new approach to protecting our food from unintentional or deliberate contamination. The U.S. Food and Drug Administration (FDA) must keep pace with these changes so that the safety of the nation's food supply remains second to none.

In the past few years, FDA has introduced several initiatives that address microbial and other food safety hazards with domestic or imported produce and that guide industry practices in the safe production of fresh-cut fruits and vegetables. FDA has also worked hard to raise awareness about food defense issues and preparedness. These are just a few things we are doing to improve food safety and food defense.

Recent nationwide recalls remind us how devastating foodborne illness can be. In the past year, contaminated peanut butter led to illnesses in more than 300 people and at least 50 hospitalizations. Contaminated spinach resulted in 206 illnesses, three deaths, and more than 100 people hospitalized. Reports of kidney failure and deaths in cats and dogs prompted a recall of more than 100 brands of pet food.

For every one of these emergencies, the FDA responded immediately to minimize harm. FDA investigators traced each problem's source and worked without delay to remove the affected products from market shelves. FDA staff continue to work diligently to protect our food supply, by containing outbreaks and preventing further illnesses.

With this FDA Food Protection Plan we are going even further. It is a forward-oriented concept that uses science and modern information technology to identify potential hazards ahead of time. By preventing most harm before it can occur, enhancing our intervention methods at key points in the food production system, and strengthening our ability to respond immediately when problems are identified, FDA can provide a food protection framework that keeps the American food supply safe.

Andrew C. von Eschenbach, M.D.
Commissioner of Food and Drugs

I. Executive Summary

FDA is implementing a Food Protection Plan (the Plan) that addresses both food safety and food defense for domestic and imported products. The Plan is integrated with the Administration's Import Safety Action Plan. The Food Protection Plan operates through a set of integrated strategies that:

- Focus on risks over a product's life cycle from production to consumption

- Target resources to achieve maximum risk reduction
- Address both unintentional and deliberate contamination
- Use science and modern technology systems

FDA's Integrated Strategy Provides Three Elements of Protection
PREVENT Foodborne Contamination • Promote Increased Corporate Responsibility to Prevent Foodborne Illnesses • Identify Food Vulnerabilities and Assess Risks • Expand the Understanding and Use of Effective Mitigation Measures
INTERVENE at Critical Points in the Food Supply Chain • Focus Inspections and Sampling Based on Risk • Enhance Risk-Based Surveillance • Improve the Detection of Food System "Signals" that Indicate Contamination
RESPOND Rapidly to Minimize Harm • Improve Immediate Response • Improve Risk Communications to the Public, Industry and Other Stakeholders

FDA recognizes the need to partner with Congress to make the changes necessary to transform the safety of the nation's food supply. This Plan identifies the administrative actions we are proposing to take within the Agency. This Plan also recommends legislative changes to strengthen FDA's ability to continue to protect Americans from foodborne illnesses.

Additional Protections that Involve Legislative Changes to FDA's Authority
PREVENT Foodborne Contamination Allow FDA to Require Preventive Controls to Prevent Intentional Adulteration by Terrorists or Criminals at Points of High Vulnerability in the Food Chain Authorize FDA to Issue Additional Preventive Controls for High-Risk Foods Require Food Facilities to Renew Their FDA Registrations Every Two Years, and Allow FDA to Modify the Registration Categories

INTERVENE at Critical Points in the Food Supply Chain
Authorize FDA to Accredit Highly Qualified Third Parties for Voluntary Food Inspections Require New Reinspection Fee From Facilities That Fail to Meet current Good Manufacturing Practices (cGM Ps) Authorize FDA to Require Electronic Import Certificates for Shipments of Designated High-Risk Products Require New Food and Animal Feed Export Certification Fee to Improve the Ability of U.S. Firms to Export Their Products Provide Parity Between Domestic and Imported Foods if FDA Inspection Access is Delayed, Limited, or Denied
RESPOND Rapidly to Minimize Harm
Empower FDA to Issue a Mandatory Recall of Food Products When Voluntary Recalls Are Not Effective Give FDA Enhanced Access to Food Records During Emergencies

FDA plans to enhance its information technology (IT) capabilities to fully support the implementation of the FDA Food Protection Plan.

II. Introduction

Every day across the country, people eat out, buy groceries, and cook meals for their families. Americans expect that all their food will be safe, and FDA plays a critical role in making sure this is true. FDA is responsible for the safety of the vast range of food Americans eat; about 80 percent of all food sold in the United States. This includes everything except for meat, poultry, and processed egg products, which are regulated by the U.S. Department of Agriculture (USDA).

In May 2007, Secretary of Health and Human Services Michael O. Leavitt and Commissioner of Food and Drugs Andrew C. von Eschenbach, M.D., charged FDA with developing a comprehensive and integrated FDA Food Protection Plan to keep the nation's food supply safe from both unintentional and deliberate contamination. Driven by science and modern information technology, the Plan aims to identify potential hazards and counter them before they can do harm. A cornerstone of this forward-thinking effort is an increased focus on prevention.

Relative rates compared with 1996-1998 baseline period of laboratory-diagnosed cases of infection with *Campylobacter*, STEC O157, *Listeria*, *Salmonella* and *Vibrio*, by year.

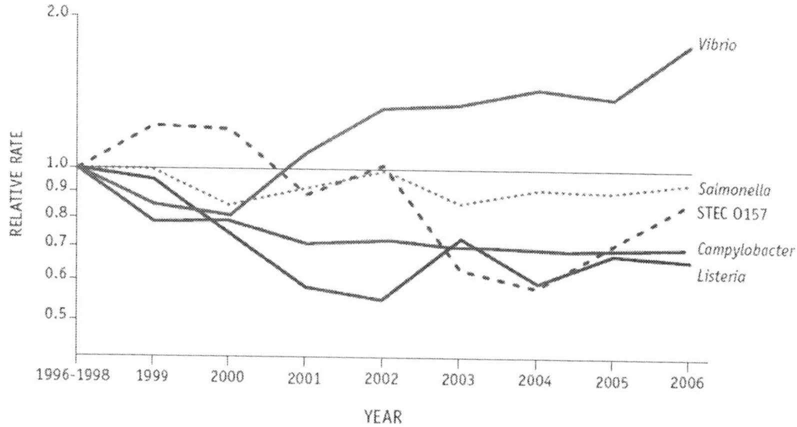

Under its FoodNet program (www.cdc.gov/foodnet), the Centers for Disease Control and Prevention (CDC) monitors foodborne microorganisms that cause illness and tracks trends. This graph shows the progress that has been made in reducing foodborne infections. Other than recent increases in Vibrio- and Shiga toxin-producing Escherichia coli (STEC) O157-related illness, the incidence of illnesses associated with these foodborne microorganisms has mostly remained steady or gone down since the late 1990s, although further progress is needed. Note that the graph represents all illnesses associated with the five types of bacteria, not just that from contaminated food. The graph also represents illnesses from foods not regulated by FDA.

The Plan builds in safety measures to address risks throughout a product's life cycle, from the time a food is produced to the time it is distributed and consumed. The Plan focuses FDA's efforts on preventing problems first, and then uses risk-based interventions to ensure preventive approaches are effective. The Plan also calls for a rapid response as soon as contaminated food or feed is detected or when there is harm to people or animals.

FDA's integrated approach, within the Food Protection Plan, encompasses three core elements: prevention, intervention and response.

- The *prevention* element means promoting increased corporate responsibility so that food problems do not occur in the first place. By comprehensively reviewing food supply vulnerabilities and developing and implementing risk reduction measures with industry and other stakeholders, FDA can best address critical weaknesses.
- The *intervention* element focuses on risk-based inspections, sampling, and surveillance at high risk points in the food supply chain. These

interventions must verify that the preventive measures are in fact being implemented, and done so correctly.

- The *response* element bolsters FDA's emergency response efforts by allowing for increased speed and efficiency. It also includes the idea of better communication with other federal, state, and local government agencies and industry during and after emergencies. Whether contamination is unintentional or deliberate, there is a need to respond quickly and to communicate clearly with consumers and other stakeholders. The communication should emphasize identifying products of concern as well as assuring the public of what is safe to consume.

FDA is committed to strengthening the nation's food protection system through implementation of the FDA Food Protection Plan. The Plan's strategic and partnered activities are driven by science and incorporate the use of 21st-century technologies.

Scope of the Food Protection Plan

1 Applies to food for people and animals
2 Addresses domestic and imported products
3 Encompasses food safety (unintentional contamination) and food defense (deliberate contamination)

FDA Regulates Roughly 80 Percent of the U.S. Food Supply

- FDA regulates $417 billion worth of domestic food and $49 billion in imported food [1] annually.
- FDA has oversight of more than 136,000 registered domestic food facilities (including more than 44,000 U.S. food manufacturers and processors and approximately 113,000 U.S. food warehouses, including storage tanks and grain elevators).[2]
- FDA or state and local authorities regulate more than 2 million farms, roughly 935,000 restaurants and institutional food service establishments, and 114,000 supermarkets, grocery stores, and other food outlets.[3] FDA provides guidance, model codes, and other technical assistance to state and local partners.
- Approximately 189,000 registered foreign facilities manufacture, process, pack, or hold food consumed by Americans.

III. Changes And Challenges

Current trends in the food industry promise better nutrition and wider choices for consumers. At the same time, multiple factors pose challenges. These include changing food production technology, patterns of human demographics and behavior, business practices, new threats, and communication issues.

Trends in Demographics and Consumption

Changes in demographics and consumption have increased consumers' susceptibility to food- borne illness. For example, by 2015, it is estimated that 20 percent of the population will be 60 or older. Older Americans are among those at highest risk for foodborne illness.

Also, the practice of a family buying a head of lettuce and preparing a salad at home is not as common. Increasingly, consumers want the convenience of opening up a bag of salad that's already prepared, and immediately serving it.It used to be that when a single head of lettuce was contaminated, the resulting illness affected one family. Now, contaminated heads of lettuce may be processed with thousands of other heads of lettuce and placed into bags of convenience salad that many consumers can buy. These bags of salad end up in thousands of homes, potentially resulting in hundreds of illnesses.

The shifting demographics have increased the numbers of susceptible consumers, and the convenience factors have meant that small problems can lead to large outbreaks–both indications of the need to make changes to ensure a continued high level of food protection.

Shifting Demographics

Our population demographics are changing. Shifting demographics means that more of the U.S. population is, and increasingly will be, susceptible to foodborne illness.

- In 2007, 20-25 percent of the population is in a high-risk category (young, older, pregnant, immune-compromised). These Americans face a risk of serious illness or death from foodborne illness*.
- In 1980, 15 percent of the population was 60 or older. By 2025, the number will be 25 percent.
- Four percent of the population is immune-compromised (transplant patients, people who are HIV positive, people receiving chemotherapy or other immunosuppressive treatments, people with chronic diseases).

Convenience Trends

- Americans are consuming more convenience foods. Foods prepared outside the home may be subject to cross-contamination from other foods, as well as contamination from food workers.
- Ready-to-eat foods (bagged salad, cut fruit) and prepared foods (including hot bars with main and side dishes, as well as salad bars) and frozen dishes that can be cooked quickly are increasing in popularity.
- Cooking in the home is decreasing—people are eating out and bringing prepared foods home.
- Spending on foodservice items, such as supermarket deli foods, accounts for about half of all U.S. food spending.

Consumption Patterns

A greater variety of foods are eaten year round. Also, foods that are consumed raw or with minimal processing are often associated with foodborne illness.

- Consumers are encouraged to make healthier food choices and increase consumption of fruits and vegetables (5-9 servings/day), including fresh produce.

---

* For example in a joint Food and Agriculture Organization of the United Nations (FAO)/World Health Organization (WHO) report on Listeria monocytogenes (LM) microbiological risk assessment, it was estimated that transplant patients had a 2,584 increased probability of becoming ill from LM, compared with a healthy adult less than 65 years old. The same report indicated that AIDS patients had an 865-fold increase and an otherwise healthy adult over the age of 65 had a 7.5-fold increase [ftp://ftp.fao.org/docrep/fao/007/y5394e/y5394e00.pdf].

- U.S. per capita consumption of fresh fruit and vegetables increased 36 percent from 1981 to 2000.
- A typical grocery store carried 173 produce items in 1987 and now carries 558 produce items.
- Produce items that were once considered seasonal are available on a year-round basis.
- Increased consumption of exotic foods whose safety hazards are not well understood.

Sources: U.S. Census Bureau and USDA Economic Research Service

Global Food Supply

There have been dramatic changes in the volume, variety, and complexity of FDA-regulated products arriving at U.S. ports. The United States trades with over 150 countries/territories with products coming into over 300 U.S. ports. In the last decade, the number of food entry lines [4] has tripled. According to the USDA Economic Research Service, approximately 15 percent of the overall U.S. food supply by volume is imported. However, in certain food categories a much higher percentage is imported. For example, approximately 60 percent of fresh fruits and vegetables consumed in the U.S. are imported, which fills the gap when U.S. domestic production is inadequate or out of season (e.g., bananas, tropical fruits, etc.). Imports of seafood rose from less than 50 percent of U.S. seafood consumption in 1980 to more than 75 percent today.

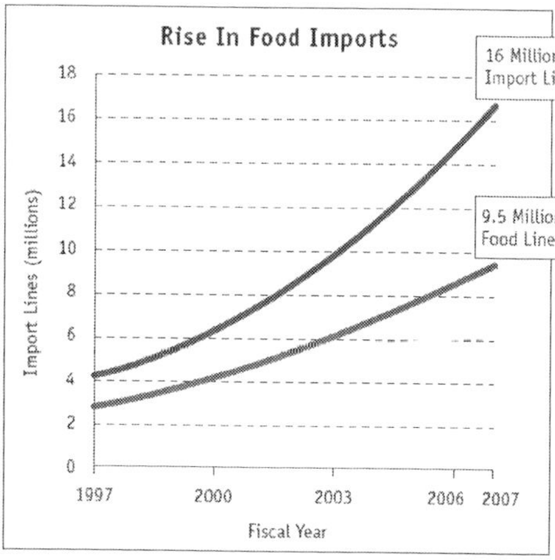

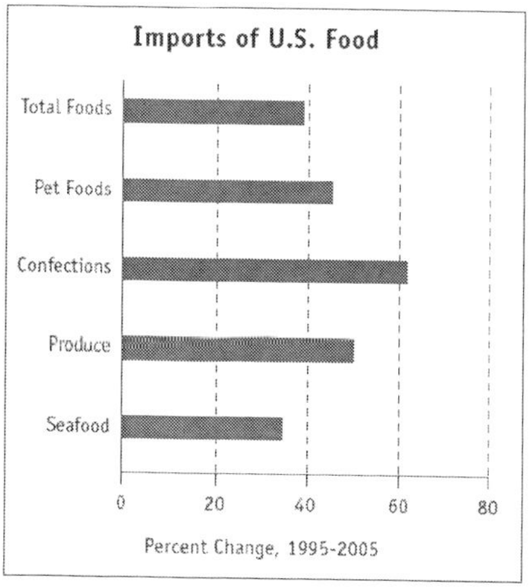

The type of imports consisted of unprocessed food ingredients with subsequent processing of those ingredients covered by FDA domestic regulatory

oversight. Today, foods that are inherently more likely to pose risks, such as ready-to-eat food products, fresh produce and seafood, account for an increasing proportion of imported foods.

This is not to suggest that food imported into the United States, as a whole, poses a greater food safety risk than domestically produced food. But increases in the volume and complexity of imported foods have taxed the limits of FDA's approach to handling imports. Currently, data on 100 percent of the shipments are submitted through the electronic systems of the U.S. Customs and Border Protection (CBP) and FDA. The data are screened electronically to determine whether the food appears to present a significant risk to public health.

Some foods are then inspected physically based on perceived risk. Food products of greater concern are physically inspected more frequently.

Currently, FDA often has very limited information regarding conditions under which most food is produced in foreign countries. While many foreign countries have well-developed regulatory systems to ensure food safety, other countries have systems that are less well- developed and that may not be able to ensure food safety to the same degree.

Growth in Foreign Manufacturers Exporting Low-Acid Canned Foods		
	1973	2004
Domestic LACF/AF Firms	742	1,300
Foreign LACF/AF Firms	34	6,700

One example of how the source of food has changed is in the import of canned or sealed fruits, vegetables, fish, and other products (collectively known as low-acid canned food/ acidified food or LACF/AF). As the table shows, the number of domestic firms nearly doubled between 1973 and 2004. By contrast, there was close to a 200-fold increase in the number of foreign firms manufacturing these products for importation into the United States during the same period.

New Threats

New Foodborne Pathogens

Symptoms of foodborne illness range from mild stomach discomfort to life-threatening neurologic, liver, and kidney syndromes. In 1999, the CDC estimated that there were around 76 million cases per year of illness from foodborne agents, with 325,000 hospitalizations and 5,000 deaths in the United States each year. These data do not identify exactly how many are spread via foods (as opposed to person-to person contact or by some other means) nor do they indicate how the food became contaminated. However, we know that the most severe cases tend to occur in people who are very young, very old, or who have compromised immune systems.

Foodborne illnesses are caused by more than 200 different foodborne pathogens (agents that can cause illness) of which we are currently aware. These include viruses, bacteria, parasites, and toxins, plus a vast number of potential chemical contaminants and metals. The variety of agents associated with foodborne illness has steadily grown over the last few decades, and there is every probability that this list will continue to increase.

One example of a newer foodborne pathogen is Enterobacter sakazakii, which can cause serious illness such as sepsis (blood infection) and meningitis (inflammation of the membrane surrounding the brain and spinal cord). In 2002, FDA, working with CDC, discovered and subsequently alerted health care professionals to clusters of E. sakazakii infections reported in a variety of locations among hospitalized newborns, particularly premature or other immuno-compromised infants who were fed powdered infant formulas.

The emergence of new foodborne pathogens requires updated technologies that can detect the presence of new agents in a variety of foods. Addressing these emerging hazards requires cooperation among industry, academia, and government to share information and establish testing protocols.

Intentional Contamination

We must also consider food as a potential vehicle for intentional contamination. Such intentional contamination of food could result in human or animal illnesses and deaths, as well as economic losses. The stark possibilities are suggested by the recent incident in which vegetable protein products, which were represented as wheat gluten and rice protein concentrate, were contaminated with melamine and melamine analogues. Though not considered an act of terrorism, the incident appeared to be a deliberate act for economic gain. It resulted in the sickness and deaths of cats and dogs, the recall of hundreds of brands of pet food products, state quarantine or voluntary holds on livestock that consumed suspect animal feed, and concern regarding the possible associated human health risks.

FDA has no reason to believe any physical harm was intended, but the melamine event indicates the danger of attempts to deliberately compromise the U.S. food system.

Pathogens Newly Associated with Foodborne Illness Since the Mid-1970's	
Campylobacter jejuni	*Campylobacter fetus*
Cryptosporidium parvum	*Cyclospora cayetanesis*
Shiga toxin-producing *E. coli*	*Listeria monocytogenes*
Noroviruses	*Salmonella* Enteritidis
Salmonella Typhimurium DT104	*Vibrio vulnificus*
Vibrio cholerae O139	*Yersinia enterocolitica*
Vibrio parahaemolyticus	*Enterobacter sakazakii*

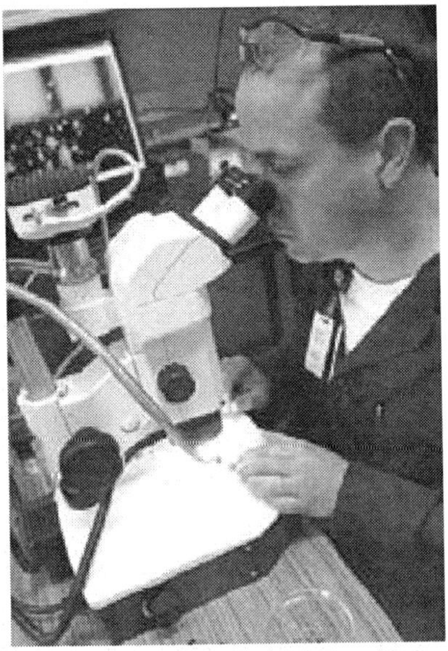

Black Star/Chris Cone for FDA

A scientist at FDA's Forensic Chemistry Center examines wheat gluten for possible melamine contamination.

Communication

Effective communication requires active collection and use of incoming information and timely communication to external groups. FDA uses the information it receives to make appropriate decisions about food safety. FDA also shares information and advice with consumers, news media, industry, and state, local, and foreign agencies. Providing information that is timely, useful, and easy to understand is critical.

FDA, states, and industry receive food safety information in various ways. Signals of potential problems come in the form of consumer complaints, inspection data, positive test results, adverse event reports, and other reports of illness. FDA is committed to improving information flow to improve detection and response to signs of trouble.

FDA collects data from several sources. Data from the testing of food, inspections, and reports of illnesses are collected in federal and state systems. Data from foodborne illness and pathogen identification are entered into systems maintained by the CDC, the lead federal agency for conducting disease surveillance and outbreak investigations. Data from imports are entered into specific import systems. Currently, states conduct 10,000 inspections under contract to FDA and another 40,000 inspections under state law. These inspections include the collection of 300,000 food samples each year.

Enabling FDA's information systems to communicate more effectively with internal and external data sources is essential. This will increase productivity of FDA staff and streamline response times during food emergencies. The overall success of the Plan depends on improving the integration and analysis of the vast amount of information collected.

Just as consumers and businesses have important roles to play in providing information to FDA, the FDA plans to improve communication with stakeholders during food emergencies. In the 2007 outbreak involving chili sauce contaminated with Clostridium botulinum, the recalled product remained on the shelves of small retailers weeks after the recall announcement. Improving outreach to all segments of the food industry will ensure that harmful products are removed from the market quickly.

IV. An Overview of the Approach

Core Elements

While American consumers enjoy one of the safest food supplies in the world, growing challenges require a new approach to food protection at FDA—an increased emphasis on prevention.

Greater attention to prevention requires closer interaction with growers, manufacturers, distributors, retailers, food service providers, and importers.

Recent outbreaks linked to fresh produce, peanut butter, and pet foods show how FDA responds quickly to contain food safety problems.

While this level of response needs to be maintained and even enhanced, there is also a need to focus more on building safety into products right from the start to meet the challenges of today. The FDA will work with the private sector to build on the actions of the food industry to ensure product safety. Building safety into products is described in one word: prevention.

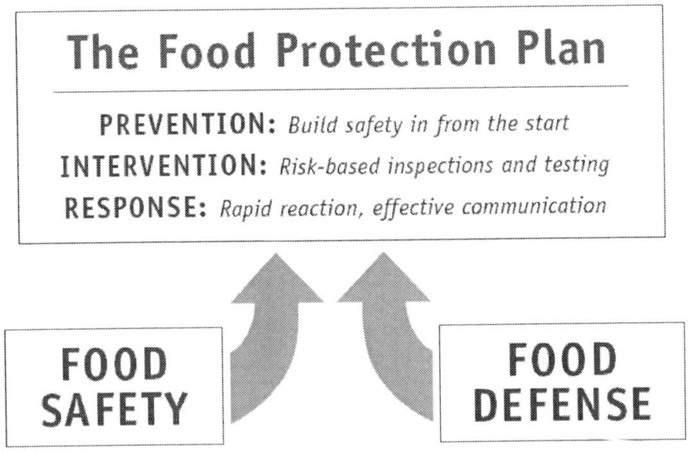

This shift to an increased emphasis on prevention is at the core of FDA's Food Protection Plan, and will be evident immediately as the FDA begins an industry-wide effort to focus attention on prevention, from general best practices for all foods to the possibility of additional measures for high-risk foods. Prevention needs to be augmented by targeted intervention that focuses inspection and testing on the areas of greatest risk. This will reduce the likelihood that contaminated products will reach consumers. However, even the best system in the world cannot prevent all incidents of foodborne illness. Along with prevention and intervention, faster and more focused response is needed once a problem is detected.

Prevention – Build Safety in from the Start.

FDA must strategically place greater emphasis on preventive measures for food safety and food defense. These measures will promote improved food protection capabilities throughout the food supply chain. This will require close interaction with growers, manufacturers, distributors, retailers and food service providers, and importers. These partners have the ability to implement preventive approaches and to require them of their suppliers. FDA will continue to work with industry, state, local, and foreign governments to further develop the tools and science needed to identify vulnerabilities and determine the most effective approaches. With regard to imports, FDA will also work with foreign governments, which have a greater ability to oversee manufacturers within their borders to ensure compliance with safety standards.

Intervention – Verify Prevention and Intervene when Risks are Identified

FDA, along with other federal agencies and state, local, and foreign governments, must undertake interventions in a coordinated and risk-based manner. Interventions, in the form of targeted inspections and testing, verify that preventive controls are working and that resources are being applied to the areas of greatest concern—either when the product is at the manufacturing facility, on its way to stores, or at a port of entry. Successful intervention will also require enhanced risk analysis, along with new detection technology to allow for faster analysis of samples. A successful and fully integrated food protection system will identify signals that indicate the need for intervention. Such signals may be a positive test for a harmful contaminant following an inspection, an industry report, a consumer complaint, or a full blown outbreak.

Response – Respond rapidly and Appropriately.

Working with its food safety partners, FDA will improve its response system to more rapidly react when signals indicate either potential or actual harm to consumers. As part of an improved response system, the FDA will develop faster and more comprehensive ways to communicate with consumers and others during a food-related emergency.

Cross-Cutting Principles

Four important cross-cutting principles will allow a comprehensive food protection approach along the entire production chain.

Principles of the Food Protection Plan
• Focus on risks over a product's life cycle from production to consumption. • Target resources to achieve maximum risk reduction. • Address both unintentional and deliberate contamination. • Use science and modern technology systems.

1. Focus on Risks over a Product's Life Cycle from Production to Consumption

Comprehensive food protection requires considering the safety and defense risks associated with foods through their whole life cycle whether domestically produced or imported. Consideration must be given to areas that are potentially vulnerable to both unintentional and intentional contamination such as the point at which food is grown or produced, every processing or manufacturing step, points involved in distribution, transport, and warehousing, as well as all the points at the retail level through distribution to consumers. It is also important to consider the role that consumers play in safeguarding food once it is in their homes.

Consideration of the risks throughout a product's life cycle is a significant shift in the Agency's approach not only for domestic products but for imported foods too. A focus on prevention at the point of manufacture based on risk will provide data to strengthen risk- based inspections domestically, at the border, and overseas. In particular, FDA plans to work with foreign governments and federal partners to ensure that foods produced in foreign facilities meet U.S. safety requirements. Risk-based targeted inspections at the border will serve as a second layer of protection, rather than the principal one.

2. Target Resources to Achieve Maximum Risk Reduction

A comprehensive risk-based approach must consider the many variables that define risk. Such variables include:

- the possibility that consuming a particular food will result in a foodborne illness due to contamination of the product, which depends on such factors as the number of microbes present or the level of a chemical or toxin present, the susceptibility of the person to the contaminating agent, and whether the food was properly handled and cooked;
- the severity of that illness, should it occur;
- the point in the production cycle where contamination is most likely to occur; and
- the likelihood of contamination and steps taken during the production cycle to reduce the possibility of contamination

Foodborne illnesses range from distressing, but tolerable, symptoms to critical and life- threatening health problems. Illness due to E. coli O157:H7 can lead to kidney failure. Exposure to botulinum toxin can cause paralysis. Other, less severe illnesses may cause diarrhea and vomiting.Some foods, such as those grown in the ground, may have little or no processing before they arrive in consumers' homes. Other foods are cooked to high temperatures (e.g., canned goods). Examining all aspects of the product life cycle helps define the areas of greatest risk. Implementation of the Plan will involve acquiring the data to best address risk, or, where the data is unavailable, working with appropriate partners to determine those risks.

3. Address Both Unintentional and Deliberate Contamination

Food safety, which traditionally refers to unintentional contamination, has been a cornerstone of public health for many years. The idea that someone may use food as a vehicle to deliberately cause harm is a risk that must be addressed. There is a heightened awareness of terrorism as a real possibility that could cause a major public health crisis. To this end, FDA has devoted significant efforts over the last six years to address food defense—defending the food supply against deliberate attack.

Whether dealing with intentional or unintentional contamination, the same regulatory experts, resources, and industry partners are involved. The best way to handle food safety and food defense is to develop approaches that appropriately address both. Although there are differences in how these events are addressed, there are also many overlaps and parallels between the two. For example, the concepts of prevention, intervention, and response apply equally to both.

Those at highest risk for serious foodborne illness include young children, older adults, pregnant women, and people with weakened immune systems.

4. Use Science and Modern Technology Systems

A successful plan for food protection is based on science. FDA's Food Protection Plan emphasizes the need to know the science underpinning how and where food becomes contaminated and the associated risks. The Plan also highlights the use of science to determine optimal interventions to reduce the likelihood of contamination. If contamination does occur, then the priority is to minimize the likelihood that it will cause significant harm. For example, successful intervention relies in large part on the science of epidemiology to understand which foods pose risks and the science of modern detection methods to identify harmful agents quickly.

The Food Protection Plan also highlights the need to further integrate information systems. Too often, sophisticated data systems lack the ability to share information. A priority in the Plan involves creating interoperable data systems, along with making current systems more interoperable, to allow for the exchange of product information along the whole life cycle. The goal is to make the most of important data from all relevant systems, and to obtain easier access to critical information.

V. THE INTEGRATED PLAN

The Food Protection Plan is based on three integrated elements of protection:

1 *Preventing* foodborne illnesses in the first place;
2 *Intervening* with risk-based FDA actions at critical points in the food supply chain; and
3 *Responding* rapidly when contaminated food or feed is detected.

Implementation of the elements will begin immediately, be phased in over time, and be integrated with the Administration's Import Safety Action Plan. All of the elements build on existing partnerships and direct resources to the areas of greatest risk.

But the FDA cannot take some key actions without new legislative authority. We summarize below in each element the new authorities needed to fully implement the Plan and strengthen our ability to protect Americans. We look forward to working productively with Congress to ensure understanding of the design of and need for these authorities.

Core Element #1: Prevention

Prevention is the first essential step for an effective, proactive food safety and defense plan. FDA's Plan implements three key prevention steps, which will move forward concurrently. The prevention steps are risk-based and will be implemented as appropriate to particular segments of the industry, taking into account that some foods are inherently safer than others.

The Plan's Key Prevention Steps
1 Promote Increased Corporate Responsibility to Prevent Foodborne Illnesses
2 Identify Food Vulnerabilities and Assess Risks
3 Expand the Understanding and Use of Effective Mitigation Measures

FDA designed its Plan for the full life cycle of food—from production to consumption whether it be domestic or imported. The prevention elements of the Plan emphasize the importance for FDA and corporations to work collaboratively to prevent food problems from occurring.

This will be accomplished through a comprehensive review of food supply vulnerabilities. FDA will work with industry and other stakeholders to develop effective tools and science to head off outbreaks of foodborne illness caused by unintentional and intentional factors.

The Food Protection Plan builds on partnerships and directs resources to the areas of greatest risk.

Some examples of enhanced corporate responsibility might include:

- evaluating safety and security vulnerabilities and possible impacts
- when appropriate, implementing preventive measures—both required and voluntary—to ensure that food is produced safely and securely
- developing a contingency plan to aid in a response in the event of contamination

1.1. Promote Increased Corporate Responsibility to Prevent Foodborne Illnesses Strengthen FDA Actions

- Meet with states and consumer groups to solicit their input on implementing preventive
- approaches to protect the food supply.
- Meet with food industry representatives to strengthen science-based voluntary
- prevention efforts, including developing best business practices and food safety
- guidelines.
- Develop written food protection guidelines for industry to a) develop food protection

- plans for produce and other food products, and b) implement other measures to
- promote corporate responsibility.
- Issue in Spring 2008, a final regulation requiring measures to prevent salmonella in shell
- eggs and resulting illnesses.
- Meet with foreign governments to share results of domestic prevention efforts and
- develop approaches for improving food safety at the source.
- Provide foreign countries with technical assistance so that they can enhance their
- regulatory systems.
- Analyze food import trend data and integrate it into a risk-based approach that focuses
- inspection resources on those imports that pose the greatest risk.
- Focus foreign inspections on high-risk firms and products.
- Improve FDA's presence overseas.

Additional Legislative Authority Needed

Allow FDA to Require Preventive Controls Against intentional Adulteration by Terrorists or Criminals at Points of High vulnerability in the Food Chain

The FDA requests authority to require entities in the food supply chain to implement measures solely intended to protect against the intentional adulteration of food by terrorists or criminals. This authority would allow FDA to issue regulations requiring companies to implement practical food defense measures at specific points in the food supply chain where intentional contamination has the greatest potential to cause serious harm, such as requiring locks on tanker trucks transporting food. The specific points would be determined using vulnerability assessments such as CARVER+Shock1, and the authority would only apply to food in bulk or batch form, prior to being packaged, which have clearly demonstrated vulnerabilities (e.g., short shelf life), and where it would affect multiple servings and there is a high likelihood of serious adverse health consequences or death from intentional adulteration. These regulations will be developed, taking into account the best available understanding of the uncertainties, risks, costs, and benefits associated with alternative options. The requirement would utilize industry best practices and would not apply to raw produce or food on farms, except for milk. FDA also proposes that firms be extended an affirmative defense in civil litigation if they comply with these controls.

Authorize FDA to issue Additional Preventive Controls for High-Risk Foods

The FDA requests explicit authority to issue regulations requiring specific types of foods (those that have been associated with repeated instances of serious health problems or death to humans or animals from unintentional contamination) be prepared, packed, and held under a system of preventive food safety controls. Such authority would strengthen the FDA's ability to require manufacturers to implement risk-based Hazard Analysis and Critical Control Point (HACCP) or equivalent processes to reduce foodborne illnesses from high-risk foods.

Require Food Facilities to Renew Their FDA Registrations Every Two years, and Allow FDA to Modify the Registration Categories

FDA requests statutory changes that would require facilities to register every two years and authorize the FDA to establish food categories within the registration system. These categories would allow FDA to tailor registration categories based on up-to-date food safety information. Under current law, FDA must use preexisting food categories that were not designed for registration purposes and therefore are of limited usefulness for evaluating potential threats to food protection. This change would ensure accurate, up-to- date registration data from facilities. Facilities whose registration remains unchanged would be able to file a simplified renewal registration or affirmation to that effect.

1The CARVER+Shock model, explained in detail at http://www. cfsan.fda.gov/ ~dms/vltcarv.html, stands for Criticality, Accessibility, Recuperability, Vulnerability, Effect, and Recognizability, plus Shock. It is available as a software tool to evaluate the potential vulnerabilities of farm-to-table supply chains of various food commodities, as well as individual facilities or processes.

Why These Actions are Important and What They Will Accomplish

Those with the biggest stake in food safety, after the consumers who eat the food, are the people and companies who grow, process, and sell food. Their livelihood depends entirely on the confidence of their customers. A poor reputation for proper food handling can drive a company to bankruptcy. Promoting increased corporate responsibility is key in shifting FDA's food protection effort to a proactive rather than a reactive one. The FDA will seek partnerships with industry to enhance consumer confidence. FDA will continue to work with industry in a) developing food protection plans that address safety and defense vulnerabilities, b) implementing prevention steps, and c) developing contingency plans to improve response to an outbreak of foodborne illness.

The FDA will primarily focus on promoting the use of risk-based, preventive systems that companies can apply at all levels of food production and processing, when appropriate. Voluntary approaches may be as basic as good manufacturing practices to ensure proper equipment sanitation and employee safety training. Potentially high-hazard food categories may require additional control measures. FDA will work with industry, consumer, and federal, state, local, and international partners to help model and promote preventive controls based on best industry practices.

FDA plans to acquire additional data to develop a better understanding of foreign country practices for food and feed. This may include the examination of best practices around the food safety control systems of other countries as well as increased understanding of the difficulties faced in implementing food protection measures. FDA will also seek to share U.S. food safety and defense best practices with foreign governments and provide technical assistance, when possible, to those countries exporting food products to the U.S. so they can enhance their regulatory systems. As part of its review of foreign systems and products, the Agency will analyze food import trend data and integrate it into a risk-based approach that focuses inspection resources on those imports that pose the greatest risk. This approach will also focus foreign inspections on high-risk firms. In the near term, a special emphasis will be placed on firms located in countries where imports into the United States have been refused repeatedly and import violations have threatened the health of U.S. consumers.

FDA's current and planned actions, along with the proposed legislative changes, would:

- Build safety and defense into the full food product life cycle–from production to consumption.
- Support work with industry, and state, local, and foreign governments to understand industry best practices and identify how and where preventive controls would work best.
- Promote the adoption of voluntary preventive controls throughout the food supply chain.
- Enhance relationships with trading partners and improve FDA's presence abroad.

1.2. Identify Food Vulnerabilities and as Asess Risks

Strengthen FDA Actions

- Work with the food industry, consumer groups, and federal, state, local and international partners to generate the additional data needed to strengthen our understanding of food safety and food defense risks and vulnerabilities.
- Use enhanced modeling capability, scientific data, and technical expertise to evaluate and
- prioritize the relative risks of specific food and animal feed agents that may be harmful.
- Establish a risk-based process to continuously evaluate which FDA-regulated products cause the greatest burden of foodborne disease.
- Work with CDC to attribute pathogens to specific foods and identify where in the production life cycle the foods became contaminated.

No additional legislative authority needed.

Why these Actions are Important and what they will Accomplish

These FDA actions provide important tools to facilitate increased corporate responsibility to prevent food contamination. These actions also address the need for additional information to better understand food safety and defense vulnerabilities and possible impacts. FDA will continue its work in this area and further engage industry and other outside groups to identify and target the greatest risks.

FDA actions will include gathering data for risk assessments and to conduct risk evaluations of commodity-agent combinations and relative risk ranking of commodities. A comprehensive, risk-based approach allows the FDA to maximize the effectiveness of its available resources by focusing on food products that have the potential to pose the greatest risk to human and animal health.

By analyzing data collected throughout the food product life cycle, we are better able to detect risks posed by food products. We are also better able to recognize key junctures where timely intervention can reduce or avoid those risks. Working with CDC, FDA will also build the capacity to attribute pathogens to specific foods and identify where in the production life cycle the foods became contaminated.

Once established and emerging risks have been identified, assessed, and ranked, we can more effectively allocate our available resources to manage these risks as addressed below.

FDA's current and planned actions would:

- Strengthen the FDA's risk assessment capabilities and capacity to provide risk evaluations efficiently and rapidly.
- Advance collaborative work with CDC, USDA, and other federal, state and local agencies to understand attribution data on the food commodities that cause foodborne illnesses.

1.3 Expand the Understanding and Use of Effective Mitigation Measures

Strengthen FDA Actions

- Focusing on higher-risk foods, develop and implement a basic research plan on sources of contamination, modes of spreading and best methods to prevent contamination.
- Research, evaluate, and develop new methods to detect food contaminants.
- Encourage outside development of new contamination detection and prevention technologies.
- Develop Web sites and other platforms for disseminating research results and new steps industry can use to address vulnerabilities.

No additional legislative authority needed.

Why these Actions are Important and what they Will Accomplish

Building on risk assessments, FDA will initiate basic research to enhance our understanding of sources of contamination, modes of spreading, and how best to prevent contamination. This information in turn will inform FDA's efforts above to promote increased corporate responsibility to implement effective preventive steps.

Focusing on higher-risk foods, FDA—working with other agencies—will undertake basic research and leverage relationships with outside organizations. The FDA will also research, evaluate, and develop new methods to detect contaminants in foods, and seek to facilitate new technologies that enhance food safety.

FDA's current and planned actions would:

- Initiate risk-driven research about sources, spread and prevention of contamination.
- Develop new mitigation tools and implement appropriate risk management strategies.

Core Element #2: Intervention

Because no plan will prevent 100 percent of food contamination, we must have targeted, risk- based interventions to provide a second layer of protection. These interventions must ensure that the preventive measures called for are implemented correctly. These interventions must also identify contaminated food that either unintentionally or intentionally circumvent our prevention plan. The Plan includes three key intervention steps.

The Plan's Key Intervention Steps

- Focus Inspections and Sampling Based on Risk
- Enhance Risk-Based Surveillance
- Improve the Detection of Food System "Signals" that Indicate Contamination

These steps emphasize targeted interventions at the point of manufacture and during distribution. They allow FDA to safeguard domestic products while increasing protection against importation of unsafe food.

Using robust risk-based analysis, FDA will conduct high-priority inspections that rely on statistical sampling and advanced risk detection tools. The FDA will verify industry busi ness practices across the food chain to ensure that effective preventive measures are in place. Gathering and analyzing test results, adverse event reports, consumer complaints, and other information will help the FDA track emerging food protection problems.

2.1. Focus Inspections and Sampling Based on Risk

Strengthen FDA Actions

- Focus food and feed safety inspections and sampling based on risk.

- Identify, evaluate and, if appropriate, validate and implement innovative foodborne pathogen detection methods and tools capable of quickly and accurately detecting contaminants in foods, such as real-time diagnostic instruments and methods that allow for rapid, on-site analysis of a particular sample.
- Train FDA and state investigators on new, technically complex, and specialized food manufacturing processes, as determined by a risk-based needs assessment, and modern inspection strategies.
- Collaborate with foreign authorities to reduce potential risk of imported food.

Additional Legislative Authority Needed

Authorize FDA to Accredit Highly Qualified Third Parties for Food Inspections

The universe of domestic and foreign food establishments subject to FDA inspection is immense and continuing to grow faster than the FDA's inspection resources. Even with the most sophisticated detection tools and laboratory capabilities, the FDA's inspection resources are finite. Therefore, legislation to authorize the FDA to accredit independent third parties, or to recognize entities that accredit, to evaluate compliance with FDA requirements would allow FDA to allocate inspection resources more effectively.

To establish such an accreditation program for voluntary food inspections, FDA would undertake a public process to determine best practices and solicit industry input in the design of the program. An FDA accreditation program would require FDA to accredit third-party organizations, or recognize an entity that accredits third parties. Third-party organizations could be, as appropriate, federal departments and agencies, state and local government agencies, foreign government agencies, or private entities without financial conflicts of interest. FDA would also:

- Audit the work of these organizations to ensure that FDA requirements were consistently assessed;
- Review their inspection reports; and
- Provide ongoing training criteria to ensure they maintain their skills and knowledge, especially as technology and requirements change over time.

FDA would use information from these accredited third-party organizations in its decision making but not be bound by such information in determining compliance with FDA requirements. Use of accredited third parties would be voluntary and might offer more in-depth review and possibly faster review times and expedited entry for imported goods manufactured in facilities inspected by accredited third parties. Use of accredited third parties may also be taken into consideration by the FDA when setting inspection and surveillance priorities.

Require New Reinspection Fee From Facilities That Fail to Meet Current Good Manufacturing Practices (cGMPs)

As part of the 2008 budget process, the Administration proposed a new user fee requiring manufacturers and laboratories to pay the full costs of reinspections and associated follow-up work when FDA reinspects facilities due to failure to meet cGMPs or other FDA requirements. Where FDA identifies violations during an inspection or issues a warning letter, FDA conducts follow-up inspections to verify a firm's corrective action. The proposed reinspection fee ensures that facilities not complying with health and safety standards bear the cost of reinspection.

Why these Actions are Important and what they will Accomplish

Effective FDA intervention means getting product risk information quickly to FDA investigators who oversee the regulated products, including a high volume of import entries. Thisinformation will allow the FDA to make better-informed decisions about what productsshould be examined more closely and tested. It also signals when to initiate further action such as additional surveillance or an enforcement action.

FDA will look to leverage the resources of outside parties to accomplish more in-depth review of food products. By improving product knowledge and communication with all of our partners, including foreign authorities and the import community, we also can identify lower-risk products requiring less FDA scrutiny at U.S. facilities and at the border. This would enable the FDA to shift more resources to evaluating more closely products that are more risky, less well known, or from unknown manufacturers.

Modern detection tools and methods are critical for effective inspections and sampling. Better detection tools will allow FDA and other partners involved in food testing to more quickly and accurately detect contaminants. Because of its relevant expertise and experience, the FDA has unique capabilities to develop these tools.

Such tools could include real-time diagnostic instruments and methods that allow for rapid, on-site analysis of a particular sample or entry, especially those that are considered high- risk. For example, rapid contamination detection technology could be expanded to cover new agents and new food types, such as produce and dairy products. This type of technology could reduce analysis time from days to minutes. Increasing the speed at which the FDA can detect problems will allow FDA to expedite import entry review decisions or provide critical health information to the public when a problem is identified.

In addition to modernizing detection tools using information technology, the FDA must modernize inspectional strategies. This means increasing the probability that investigators will observe and identify potential problems.

FDA's current and planned actions, along with the proposed legislative changes, would result in:

- Focused risk-based inspections and sampling across the food chain.
- Development of rapid detection and testing tools.
- Increased involvement of federal, state, local, and foreign governments, in coordination with other food safety partners.
- Greater product knowledge and oversight through the accreditation of independent third parties.
- Modernized inspectional strategies.

2.2. Enhance Risk-Based Surveillance

Strengthen FDA Actions

- Further enhance FDA's ability to target imported foods for inspection based on risk and publish the *Prior Notice of Imported Foods* Final Rule in 2008 as part of Bioterrorism Act implementation.
- Conduct foreign food and animal feed inspections more efficiently using the tools designed to target high-risk firms.
- Use advanced screening technology at the border.
- Improve data quality and handling capacity for food imports.
- Enhance information sharing agreements with key foreign countries.

Additional Legislative Authority Needed

Authorize FDA to Require Electronic import Certificates for Shipments of Designated High-Risk Products

For food imports, the burden falls primarily on FDA to inspect and detect contamination at the U.S. border. With the explosion in import volume, this burden has become a serious challenge. The FDA should have the option of moving the inspection of high-risk products of concern "upstream" by entering into agreements with the exporting country's regulatory authority for that authority (or an FDA-recognized third-party inspector) to certify each shipment or class of shipments for compliance with FDA's standards prior to shipment. FDA would apply this requirement for imported products that have been shown to pose a threat to public health for U.S. consumers and thus would be unlike other imports where there is no such showing of risk. Such import certificate programs would be used for designated products imported from countries with whom FDA has concluded an agreement on a certification program that provides a level of safety sufficient to meet HHS/FDA standards. FDA would implement the government-to-government agreement by requiring importers to provide certificates from either relevant government agencies or accredited third parties.

While FDA would retain the authority to verify the safety of imported products, this approach shares the burden of ensuring the safety of food products with the exporting country. Shipments that fail to meet requirements would be refused entry.

For such a system to be effective, FDA will have to establish an in-depth collaboration with the relevant foreign government authority to ensure that the standards, processes, and criteria the foreign authority or third party uses in certifying products are sufficient to ensure compliance with FDA food safety standards. The FDA will also have to take several steps to ensure a secure system that prevents counterfeiting of the certificates and takes into consideration transshipment of products as a way to avoid certification.

FDA would use non-discriminatory science and risk-based criteria to determine the focus of this proposed authority and would use the authority only to the extent necessary to protect human or animal life or health.

Require New Food and Animal Feed Export Certification Fee to improve the Ability of U.S. Firms to Export Their Products

As part of the 2008 budget process, the Administration proposed a new export certification fee for the issuance of export certificates for foods and feeds to those situations where exportation is restricted without this type of certificate. Private sector exporters would bear the cost of the program, but would reap its benefits through the FDA's enhanced ability to facilitate product exports. Importantly, collection of these user fees will enable the FDA to issue certificates without redirecting resources from other critical food and animal feed safety programs devoted to protecting the public health. Such fees are currently collected by the FDA for export certificates for drugs and devices.

Provide Parity Between Domestic and imported Foods if FDA inspection Access is Delayed, Limited, or Denied

While FDA currently has the authority to obtain a warrant or initiate criminal proceedings if it is denied access to inspect facilities here in the U.S., its ability, under the Federal Food, Drug and Cosmetic Act, to enforce the inspection provisions for overseas sites is very limited. In particular, the FDA cannot refuse admission of food, even if its efforts to conduct a foreign inspection were unduly delayed, limited or denied at a facility where the product was manufactured, processed, packed or held. Having the authority to prevent entry of food from firms that fail to provide FDA access will enable the FDA to keep possibly unsafe food from entering U.S. markets. This authority provides strong motivation for firms to allow FDA to perform inspections, motivation similar to that provided to domestic firms. The authority would include several procedural safeguards, including an informal hearing if food is refused admission into the United States, such as is available for food that may be refused entry for other reasons.

Sampling the highest priority imports, especially those posing a significant public health threat, is critical ...

Why these Actions are Important and what they will Accomplish

FDA must prevent products that pose food safety and food defense threats from entering the United States. A targeted, risk-based approach to foreign product regulation is essential. Sampling the highest priority imports, especially those posing a significant public health threat, is critical and dependent on data related to the practices in the foreign facility. The activity will enhance FDA's import programs and focus these programs on the life cycle of the imported product, through such means as enhanced use of information-sharing agreements with key foreign countries.

In addition, FDA will continue to look for enhanced ways to use risk-based screening technology to identify products that pose health risks at the border. For example, a screening technology prototype is currently being tested on imported seafood products in Los Angeles. If demonstrated successful, this technology could be extended to other imported products and ports, thus enhancing the FDA's ability to quickly screen products at the border.

FDA's current and planned actions, along with the proposed legislative changes, would:

- Better focus on the imported products' total life cycle.
- Improve data systems to monitor foreign-produced food products.

2.3 Improve the Detection of Food System "Signals" that Indicate Contamination

Strengthen FDA Actions

- Deploy new rapid screening tools and methods to identify pathogens and other contaminants.
- Improve FDA's adverse event and consumer complaint reporting systems, including capturing complaints made to food manufacturers and distributors.
- Work to create a Reportable Food Registry for reports of a determination that there is a reasonable probability that the use of or exposure to an article of food will cause serious harm or death to humans or animals [as defined in the 2007 Food and Drug Administration Amendments Act (FDAAA)]. Under FDAAA, industry is expected to report such situations to the FDA within 24 hours.
- Work to create an Early Warning Surveillance and Notification System to identify adulterated pet food products, outbreaks of pet illness and to provide notice to veterinarians and other stakeholders during pet food recalls (as defined in the 2007 Food and Drug Administration Amendments Act or FDAAA).

No additional legislative authority needed.

Why these Actions are Important and what they will Accomplish

FDA can better detect and more quickly identify risk "signals" in the food supply chain via two key approaches: 1) deploying new rapid screening tools and methods to identify pathogens and other contaminants; and 2) enhancing its ability to "map" or trace adverse events back to their causes (whether reported to FDA or the food manufacturer or distributor) by improving its adverse event and consumer complaint reporting systems. This additional information will serve as a supplemental warning indicator for trending emerging food protection problems.

To provide the information necessary to allow for early detection of, and intervention with, contaminated animal feed, FDA will develop a centralized database for veterinarians that captures data on food safety incidents and the causes of food-related illness. The FDA will populate the database with key information from the veterinary community, veterinary hospitals, and other private U.S. sources.

FDA's current and planned actions would identify:

signals that may indicate a problem with food from routine testing, consumer complaints, industry reporting and documented illnesses.

Core Element #3: Response

During the past year, FDA responded to food safety problems with contaminated spinach, lettuce, vegetable proteins, and peanut butter, among other foods. Whether contamination is unintentional or deliberate, there is a need to respond faster and communicate more effectively with consumers and other partners.

The following key response steps will increase FDA's ability to quickly identify food safety problems, better coordinate a rapid emergency response among FDA, state and local government response teams as appropriate, and improve communications to the public, industry and other partners. This will better protect public health, help reduce the economic hardship affected industries face, and most importantly, maintain consumer confidence in the U.S. food supply following an incident.

Improve Immediate Response

Strengthen FDA Actions

- Enhance the data collection, incident reporting and emergency response mapping capabilities of FDA's Emergency Operations Network Incident Management System.

- Work with stakeholders to develop an action plan for implementing more effective trace-back process improvements and technologies to more rapidly and precisely track the origin and destination of contaminated foods, feed, and ingredients.
- Increase collaboration with foreign, federal, state, and local FDA partners to identify a contamination source, remove contaminated products, and implement corrective actions.
- Work with CDC and other selected federal, state, and local testing labs to communicate real-time testing results among FDA and lab members.

Additional Legislative Authority Needed

Empower FDA to issue a Mandatory Recall of Food Products when voluntary Recalls are not Effective

Although FDA has the authority to seize adulterated or misbranded food, this is not a practical option when contaminated product has already been distributed to hundreds or thousands of locations. And while the FDA has been able to accomplish most recalls through voluntary actions by product manufacturers or distributors, there are situations in which firms are unwilling to conduct a recall. In such situations FDA needs the ability to require a firm to conduct a recall to ensure the prompt and complete removal of food from distribution channels. This authority would be limited to foods that the Secretary has reason to believe are adulterated and present a threat of serious adverse health consequences or death. It would be imposed only if a firm refuses or unduly delays conducting a voluntary recall. An order to recall food could only be issued by the HHS Secretary, Deputy Secretary, or Commissioner of Food and Drugs, and would be accompanied by appropriate due process rights.

Provide FDA Enhanced Access to Food Records During Emergencies

During food-related emergencies, the FDA needs more complete and streamlined access to records necessary to identify the source of foodborne illness and take needed action. Improved access to information, including records related to an article of food or related articles of food that may present a threat, will enhance FDA's ability to identify problems, respond quickly and appropriately, and protect public health.

Currently, emergency access to records is limited to instances where, for an article of food, FDA has a reasonable belief that the food is adulterated and presents a threat of serious adverse health consequences or death. FDA proposes to expand access to records of related articles of food, such as food produced on the same manufacturing line. FDA also proposes, in food-related emergencies, to remove the adulteration requirement to allow its inspectors access to records in emergency situations where FDA has a reasonable belief that an article of food presents a threat of serious adverse health consequences or death. The recent melamine situation in which FDA had early clinical evidence that a specific food was causing illness in pets but did not have clear evidence of a specific adulteration is an example of such a scenario.

The records access would relate only to safety or security of the food and would not apply to records pertaining to recipes, financial data, pricing data, personnel data, research data, and sales data. The requirement would not impose any new recordkeeping burdens, and would maintain the current statutory exclusions for the records of farms and restaurants.

As part of that effort, FDA will work with selected federal, state, and local testing labs to communicate real-time testing results among FDA and lab members.

FDA will also increase collaboration with foreign, state, and local regulators to identify the source of contamination, remove contaminated products as quickly as possible, and implement measures needed to prevent future contamination.

These improvements will allow FDA to quickly isolate problems, prevent contaminated products from reaching consumers, and ensure targeted recalls of products. Such steps aim to minimize the public health and economic impact from an outbreak.

FDA's current and planned actions, along with the proposed legislative changes, would:

- Enhance the nation's food emergency response system.
- Expand the FDA's trace-back process.
- Improve multi-partner collaborations, including with foreign regulators.

3.2. Improve Risk Communications to the Public, Industry, and other Stakeholders

Strengthen FDA Actions

- Work with communications and media experts, including FDA's Risk Communication Advisory Committee, to design and conduct consumer communications and behavior response studies.
- Update the Food Protection Risk Communications Plan using the most effective strategies for sharing information with consumers.
- Build a consumer Web site to communicate relevant food protection information.
- In a food-related emergency, implement this communications plan, including utilizing all relevant media and technologies to reach consumers, retailers, industry, public health officials, and other stakeholders resulting in a better informed and thus more resilient population.

No additional legislative authority needed.

Why these Actions are Important and what they will Accomplish

Consumers protect themselves and their families from foodborne illness by responding promptly to FDA alerts. Important messages must be communicated clearly and through multiple forms of media to be effective, because different segments of the population use different technologies, ranging from television and newspapers to text messages and podcasts. In addition, major segments of the population do not use English as their primary language and rely on still other sources of information. This increases the challenge of implementing effective communication strategies.

Retailers, public health officials, industry and other key stakeholders likewise use an array of communications vehicles and sources. FDA's communication strategy during emergencies must use all such media to reach these different audiences and ensure that potentially harmful products are removed promptly.

FDA will enhance its risk communication program through aggressive, targeted food safety campaigns that disseminate clear and effective messages and regular updates through multiple venues to all targeted audiences.

VI. Enhance Information Technology

In support of all three components of the Food Protection Plan, FDA plans to enhance its IT systems related to both domestic and imported foods. The focus will be to help the FDA more rapidly identify food importers, and maintain, update, and search records on food facilities and shipments more efficiently.

In particular, FDA will enhance collaboration with CBP on IT systems to more accurately identify firms involved in the food import supply chain during the import screening and review processes. These systems will allow for analysis of historical risk data about firms when making entry decisions for the firms' products.

A new systems approach can eliminate many problems with our current data. For example, assigning a unique identifier will eliminate duplicate records and make risk data about a firm easier to access. Policies for requiring the use of the new single national identifier will need to be established and agreed upon, recognizing the impact on industry worldwide.

Nearly all FDA business processes will benefit from more reliable and accurate information. Implementation of a new system will require a coordinated multi-agency effort that will benefit all federal agencies that process imported foods. CBP's existing data and ongoing activity will play a key role.

Finally, FDA will ensure that its infrastructure and disaster recovery system for IT systems and data are ready to deal with planned (maintenance and upgrades) and unplanned outages. This will provide the necessary support for import operations, which require the availability of multiple FDA systems around the clock. As an example, shipments arrive at U.S. ports day and night, and Prior Notice data are submitted at all hours. IT systems provide screening of the data as they are submitted, and Prior Notice Center (PNC) staff work around the clock to review the risk presented by shipments before their arrival. The PNC needs to review shipment data in as little as two hours from submission. Any interruption in the availability of the computer systems prevents the filing and timely review of information. This affects the flow of goods into the United States, and poses a safety risk to consumers.

An integrated, IT infrastructure—with data gathering, sorting, mining, and trending capability built into the systems—is critical to the success of FDA's food protection efforts.

The Public Health Impact of the Food Protection Plan

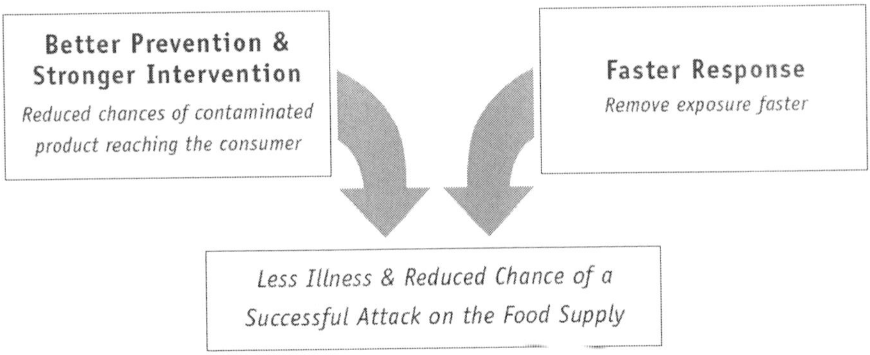

VII. CONCLUSION

Ensuring that FDA-regulated products are safe and secure is a vital part of FDA's mission— to protect and promote public health. The FDA remains committed to working closely with its partners to protect the nation's food supply.

In the United States, market forces give companies a strong motivation to be vigilant and even innovative in ensuring food safety. The laws of regulation must encourage, not disrupt, these motivations. Rather than taking over responsibility from food companies, FDA wants to protect their flexibility to pursue it vigorously.Although we have made progress, much remains to be done. Recent incidents of contaminated food and animal feed have highlighted the importance of a strong food protection system. Americans rightly expect to purchase food without having to worry about safety.

Rising food imports, increasing consumption of convenience foods, and new foodborne pathogens are among the challenges we face. To address these challenges, we must move toward a food safety and defense system that is more proactive and strategic.

FDA's Food Protection Plan contains three core elements—prevention, intervention, and response—with greater emphasis on preventive measures that keep contaminated food from ever reaching consumers. The Plan operates through a set of integrated strategies that address the product life cycle, a risk-based allocation of resources, the integration of food safety and food defense, and builds on a foundation of science and modern information systems.

FDA's Food Protection Plan complements the nation's strategic framework for import safety, which was released by the U.S. Department of Health and Human Services in September 2007. Both plans focus efforts on working smarter and better with importers, manufacturers, and other government agencies.

FDA will aggressively pursue the Food Protection Plan so that U.S. consumers can be assured that their food remains among the safest in the world.

REFERENCES

[1] Based on FDA value-of-shipment information, 2003.

[2] Facilities that are engaged in more than one type of activity (e.g., manufacturing and warehousing) are counted in both categories; thus, the sum of the individual numbers of type of facilities exceeds the number of total registered facilities.

[3] Data from U.S. Department of Agriculture, National Restaurant Association, and U.S. Census Bureau.

[4] An entry line means each portion of an import shipment that is listed as a separate item on an entry document. Items in an import entry having different tariff descriptions must be listed separately.

In: Food Protection and Safety
Editor: Lyman F. Thomasson

ISBN 978-1-60692-187-6

Chapter 2

FEDERAL OVERSIGHT OF FOOD SAFETY. FDA'S FOOD PROTECTION PLAN PROPOSES POSITIVE FIRST STEPS, BUT CAPACITY TO CARRY THEM OUT IS CRITICAL*

Lisa Shames

WHY GAO DID THIS STUDY[1]

The Food and Drug Administration (FDA) is responsible for ensuring the safety of roughly 80 percent of the U.S. food supply, including $417 billion worth of domestic food and $49 billion in imported food annually. The recent outbreaks of E. coli in spinach, Salmonella in peanut butter, and contamination in pet food highlight the risks posed by the accidental contamination of FDA-regulated food products. Changing demographics and consumption patterns underscore the urgency for effective food safety oversight. In response to these challenges, in November 2007, FDA and others released plans that discuss the oversight of food safety. FDA's Food Protection Plan sets a framework for food safety oversight. In addition, FDA's Science Board released FDA Science and Mission at Risk, which

* Excerpted from GAO-08-435T, dated January 29, 2008.

concluded that FDA does not have the capacity to ensure the safety of the nation's food supply.

This testimony focuses on (1) federal oversight of food safety as a high-risk area that needs a governmentwide reexamination,

(2) FDA's opportunities to better leverage its resources, (3) FDA's Food Protection Plan, and (4) tools that can help agencies to address management challenges. To address these issues, GAO interviewed FDA officials; evaluated the Food Protection Plan using a GAO guide for assessing agencies' performance plans; and reviewed pertinent statutes and reports. GAO also analyzed data on FDA inspections and resources.

What GAO Found

FDA is one of 15 agencies that collectively administer at least 30 laws related to food safety. This fragmentation is the key reason GAO added the federal oversight of food safety to its High-Risk Series in January 2007 and called for a governmentwide reexamination of the food safety system. We have reported on problems with this system—including inconsistent oversight, ineffective coordination, and inefficient use of resources.

FDA has opportunities to better leverage its resources. Efficient use of resources is particularly important at FDA because we found that its food safety workload has increased in the past decade, while its food safety staff and funding have not kept pace. GAO has recommended that FDA establish equivalence agreements with other countries to shift some oversight responsibility to foreign governments, explore the potential for certifying third party inspections, and consider accrediting private laboratories to inspect seafood, among other actions. We also reported that FDA and the U.S. Department of Agriculture (USDA) conduct similar inspections at 1,451 facilities that produce foods regulated by both agencies. To reduce overlaps, we recommended that, if cost-effective, FDA enter into an agreement to commission USDA inspectors at such facilities. FDA incorporated some of these recommendations in its Food Protection Plan.

FDA's Food Protection Plan also proposes some positive first steps intended to enhance its oversight of food safety. Specifically, FDA requests authority to order food safety recalls and issue additional preventive controls for high-risk foods, both of which GAO has previously recommended. However, more specific information about its strategies and the resources FDA needs to implement the plan would facilitate congressional oversight. FDA officials acknowledge that implementing the Food Protection Plan will require additional resources. Without a clear description of resources and strategies, it will be difficult for Congress to assess the likelihood of the plan's success in achieving its intended results.

The Science Board cites numerous management challenges that have contributed to FDA's inability to fulfill its mission, including a lack of a coherent structure and vision, insufficient capacity in risk assessment, and inadequate human capital recruitment and retention. In light of these challenges, GAO has identified through other work some tools that can help agencies improve their performance over time. For example, a Chief Operating Officer/Chief Management Officer can help an agency address longstanding management problems that are undermining its ability to accomplish its mission and achieve results. In addition, a well-designed commission can produce specific practical recommendations that Congress can enact. Critical success factors that can help ensure a commission's success include a statutory basis with adequate authority, a clear purpose and timeframe, leadership support, an open process, a balanced membership, accountability, and resources.

Mr. Chairman and Members of the Subcommittee:

I am pleased to be here today to discuss the resources the Food and Drug Administration (FDA) uses to meet one of its key regulatory responsibilities, the oversight of food safety. FDA is responsible for ensuring the safety of roughly 80 percent of the U.S. food supply, including $417 billion worth of domestic food and $49 billion in imported food annually. Contaminated food can harm human health, have severe economic consequences, and undermine consumer confidence in the government's ability to ensure the safety of the U.S. food supply. The recent outbreaks of E. coli in spinach, Salmonella in peanut butter, and contamination in pet food, highlight the risks posed by the accidental contamination of FDA-regulated food products. For example, according to FDA, the recent California spinach E. coli outbreak resulted in 205 confirmed illnesses and 3 deaths, and industry representatives estimate that economic losses ranged from $37 million to $74 million.

Changing demographics and consumption patterns underscore the urgency for effective food safety oversight. According to FDA, shifting demographics mean that more of the U.S. population is, and increasingly will be, susceptible to foodborne illnesses. The risk of severe and life- threatening symptoms from infections caused by foodborne pathogens is higher for older adults, young children, pregnant women, and immune compromised individuals. According to FDA, these groups make up about 20 to 25 percent of the U.S. population. In addition, we are increasingly eating foods that are consumed raw or with minimal processing and often associated with foodborne illness. For example, according to the U.S. Department of Agriculture (USDA), leafy greens such as spinach, are the category of produce most likely to be associated with an outbreak, and the average consumer ate 2.4 pounds of fresh spinach in 2005—a 180 percent increase over 1992.

In response to these increasing challenges, FDA and other agencies recently released plans that discuss the oversight of food safety. In November 2007, FDA released its Food Protection Plan, which sets forth FDA's framework for overseeing the safety of food.[2] Concurrently, a twelve-agency working group presented to the President its Action Plan for Import Safety,[3] which contains, among other things, recommendations for improving the safety of food imports entering the United States. Both plans spell out numerous actions FDA plans to take to enhance food safety, including writing new food protection guidelines for industry and helping foreign countries improve their regulatory systems. The plans also request new legislative authorities. One requested legislative authority is for enhanced access to a food company's records during food safety emergencies. Subsequently, FDA's Science Board, an advisory board to the agency, released a report titled, FDA Science and Mission at Risk.[4] This report, which is the focus of today's hearing, concluded that FDA is not positioned to meet current or emerging regulatory needs, and stated that FDA does not have the capacity, such as staffing and technology, to ensure the safety of the nation's food supply. In addition, the report found that FDA's ability to provide its basic food system inspection, enforcement, and rulemaking functions is severely eroded, as is its ability to respond to outbreaks of foodborne illnesses in a timely manner and to develop and keep pace with the science needed to prevent food safety problems. The report stated that the system cannot be fixed using available resources, and its primary food safety recommendation was that FDA needs additional resources to fulfill its regulatory mandate.

I will focus on four key points: (1) federal oversight of food safety is a high-risk area that needs a governmentwide reexamination, (2) FDA has opportunities to better leverage its resources, (3) FDA's Food Protection Plan proposes some positive first steps but additional information on the plan's strategies and resources can facilitate congressional oversight, and (4) tools such as a commission or chief operating officer can help agencies to address management challenges. This testimony is based on new and previously issued work. Today, GAO is also testifying on another FDA regulatory responsibility—inspections of medical device manufacturers.[5] These and other recent testimonies on food and drug safety offer observations on FDA's management capacity.

To assess FDA's Food Protection Plan, we interviewed FDA officials; reviewed pertinent statutes and reports; and evaluated the plan using a GAO guide for assessing agencies' performance plans. To analyze data on FDA inspections, we examined data from FDA and determined that they were sufficiently reliable for our analyses. We also reviewed funding data from the Science Board and analyzed the data in real terms. To provide updated information on our previously issued reports, we gathered information on the status of our recommendations. We conducted our work in January 2008 in accordance with generally accepted government auditing standards. Those standards require that we plan and perform the audit to obtain sufficient, appropriate evidence to provide a reasonable basis for our findings and conclusions based on our audit objectives. We believe that the evidence obtained provides a reasonable basis for our findings and conclusions based on our audit objectives.

FEDERAL OVERSIGHT OF FOOD SAFETYIS A HIGH-RISK AREA THAT NEEDS A GOVERNMENTWIDE REEXAMINATION

While part of today's hearing focuses specifically on FDA's responsibilities for the oversight of food safety, it is important to note that FDA is one of 15 federal agencies that collectively administer at least 30 laws related to food safety. This fragmentation is a key reason we designated federal oversight of food safety as a high-risk area. Two agencies have primary responsibility—FDA is responsible for the safety of virtually all foods except for meat, poultry, and processed egg products, which are the responsibility of USDA. In addition, among other agencies, the National Marine Fisheries Service (NMFS) in the Department of Commerce conducts voluntary, fee-for-service inspections of seafood safety and quality; the Environmental Protection Agency (EPA) regulates the use of pesticides and maximum allowable residue levels on food commodities and animal feed; and the Department of Homeland Security is responsible for coordinating agencies' food security activities. This federal regulatory system for food safety, like many other federal programs and policies, evolved piecemeal, typically in response to particular health threats or economic crises.

In January 2007, we added the federal oversight of food safety to our High-Risk Series,[6] which is intended to raise the priority and visibility of government programs that are in need of broad-based transformation to achieve greater economy, efficiency, effectiveness, accountability, and sustainability. Over the past 30 years, we have reported on issues—for example, the need to transform the federal oversight framework to reduce risks to public health as well as the economy—that suggest that the federal oversight of food safety could be designated as a high-risk area. The fragmented nature of the federal food oversight system calls into question whether the government can plan more strategically to inspect food production processes, identify and react more quickly to outbreaks of foodborne illnesses, and focus on promoting the safety and integrity of the nation's food supply.

While we have reported on problems with the federal food safety system—including inconsistent oversight, ineffective coordination, and inefficient use of resources—most noteworthy for today's hearing is that federal expenditures for the oversight of food safety have not been commensurate with the volume of foods regulated by the agencies or consumed by the public. We have reported that four agencies—USDA, FDA, EPA, and NMFS—spent a total of $1.7 billion on food safety-related activities in fiscal year 2003.[7] USDA and FDA were responsible for nearly 90 percent of those federal expenditures. However, the majority of federal expenditures for food safety inspection were directed toward USDA's programs for ensuring the safety of meat, poultry, and egg products even though USDA is responsible for regulating only about 20 percent of the food supply. In contrast, FDA accounted for only 24 percent of expenditures even though it is responsible for regulating about 80 percent of the food supply.

Others have called for fundamental changes to the federal food safety system overall. In 1998, the National Academy of Sciences concluded that the system is not well equipped to meet emerging challenges.[8] In response to the Academy's report, the President established a Council on Food Safety which released a Food Safety Strategic Plan in January 2001. The plan recognized the need for a comprehensive food safety statute and concluded, "the current organizational structure makes it more difficult to achieve future improvements in efficiency, efficacy, and allocation of resources based on risk."

While many of the recommendations we made have been acted upon, a fundamental reexamination of the federal food safety system is warranted.

Taken as a whole, our work indicates that Congress and the executive branch can and should create the environment needed to look across the activities of individual programs within specific agencies, including FDA, and toward the goals that the federal government is trying to achieve. To that end, we have recommended, among other things, that Congress enact comprehensive, uniform, and risk-based food safety legislation and commission the National Academy of Sciences or a blue ribbon panel to conduct a detailed analysis of alternative organizational food safety structures.[9] We have also recommended that the executive branch reconvene the President's Council on Food Safety to facilitate interagency coordination on food safety regulation and programs. According to documents on the council's Web site, the current administration has not reconvened the council.

These actions can begin to address the fragmentation in the federal oversight of food safety. Going forward, to build a sustained focus on the safety and integrity of the nation's food supply, Congress and the executive branch can integrate various expectations for food safety with congressional oversight and through agencies' strategic planning processes, including FDA's. We have previously reported that the development of a governmentwide performance plan that is mission- based, is results-oriented, and provides a cross-agency perspective offers a framework to help ensure agencies' goals are complementary and mutually reinforcing. Further, with pressing fiscal challenges, this plan can help decision makers balance trade-offs and compare performance when resource allocation and restructuring decisions are made.

FDA HAS OPPORTUNITIES TO BETTER LEVERAGE ITS RESOURCES

In response to the nation's fiscal challenges, agencies may have to explore new approaches to achieve their missions, and we have identified options for FDA to better leverage its resources. Efficient use of resources is particularly important at FDA because, while its food safety workload has increased in the past decade, resources have not kept pace. FDA has proposed actions toward implementing some of these options. Our analysis of FDA data shows that while FDA received increased funding for new bioterrorism-related responsibilities in 2003, subsequent staffing levels and funding have not kept pace with the agency's growing responsibilities. Specifically, the number of FDA-regulated domestic food establishments increased more than 10 percent from fiscal years 2003 to 2007—from about 58,260 in 2003 to about 65,520 in 2007. Additionally, FDA notes that there have been dramatic changes in the volume, variety, and complexity of FDA-regulated products arriving at U.S. ports, and recently reported that the number of food import entry lines has tripled in the past ten years.[10] Meanwhile, staffing for FDA's Center for Food Safety and Applied Nutrition (CFSAN) has decreased. According to the Science Board, the number of staff years for CFSAN operations at headquarters dropped about 14 percent, from 950 in fiscal year 2003 to 812 in fiscal year 2006. During that same time period, field-based staff responsible for carrying out inspection and enforcement activities for CFSAN-regulated products dropped by 255 staff years, or about 11.5 percent—from 2,217 in fiscal year 2003 to 1,962 in fiscal year 2006. In addition, while CFSANrelated funding at headquarters and in the field increased from $407 million in fiscal year 2003 to $439 million in fiscal year 2006, this represents a decrease in real terms from about $457 million to about $451 million during that period. One consequence is that foreign inspections have declined: GAO analysis of FDA data shows that inspections of foreign food firms, which number almost 190,000, decreased from 211 in fiscal year 2001 to fewer than 100 in fiscal year 2007. The Science Board considered the funding issues to be more acute for CFSAN than for other FDA programs: unlike the FDA programs responsible for drugs, biologics, and medical devices, which charge manufacturers hundreds of millions of dollars in user fees each year, CFSAN is not authorized to charge user fees for its services.

Recent GAO work has identified opportunities for FDA to better leverage its resources. Specifically, in 2004 we reviewed FDA's imported seafood safety program and identified several options that FDA could consider to augment its resources and enhance its current program.[11] We found that FDA's seafood safety program had shown some progress from a 2001 review. For example, FDA increased its laboratory testing of seafood products at ports of entry from less than 1.0 percent in fiscal year 1999 to about 1.2 percent in fiscal year 2002. We also recommended several options for enhancing FDA's oversight of seafood while leveraging outside resources. Some of these options are presented in FDA's Food Protection Plan. We recommended that FDA:

- Make it a priority to establish equivalence agreements with other countries. Subject to its jurisdiction, FDA could certify that countries exporting food products to the United States have equivalent food safety systems before food products from those countries can enter the United States. Such agreements would shift some of FDA's oversight burden to foreign governments. While FDA has not yet established equivalence agreements with any foreign countries, the Food Protection Plan requests that Congress allow the agency to enter into agreements with exporting countries to certify that foreign producers' shipments of designated high-risk products comply with FDA standards.
- Explore the potential for certifying third-party inspectors. FDA could consider developing a program that uses certified third-party inspectors to conduct inspections on its behalf, both at foreign processing firms and domestic importers of seafood. FDA's Food Protection Plan requests

authority from Congress to accredit third parties to conduct voluntary inspections for foods, and FDA officials told us that they envision using third-party inspectors to inspect foreign facilities, where FDA conducts few inspections. If FDA receives this authority, it can take lessons from its own implementation of third-party inspection programs for medical device manufacturing establishments. As we are reporting in a separate statement today, few inspections of these establishments have been conducted through FDA's two accredited third-party inspection programs.

- Consider accrediting private laboratories to test seafood. Currently, FDA does not accredit or use any private laboratories to collect or analyze seafood samples. However, for some seafood violations, it allows seafood firms to use private laboratories to provide evidence that imported seafood previously detained because of safety concerns is now safe and can be removed from the detention list at the port of entry. We recommended that FDA consider accrediting private laboratories because it could leverage outside resources while providing FDA greater assurance about the quality of the laboratories importers use to demonstrate that their products are safe. FDA has not formally changed its policies or practices, but the Action Plan for Import Safety notes that FDA intends to issue guidance by mid-2008 on sampling and testing of imported products, including the use of accredited private laboratories submitting data to FDA on food safety.
- Develop a memorandum of understanding with the National Oceanic and Atmospheric Administration (NOAA) to use NOAA's Seafood Inspection Program resources to complete inspections on FDA's behalf. NOAA officials said that they could provide various services to augment FDA's regulatory program for imported seafood, including inspection, training, and product sampling services. FDA has been working on a program to refer certain export-related work to NOAA, and it is in discussions with NOAA about commissioning its inspectors, but to date, nothing is finalized or operational.

We have not reviewed these actions to determine whether they adequately address our recommendations.

We separately reported on overlaps we identified in the federal oversight of food safety, such as overlapping inspection and training activities that exist among the agencies conducting food safety functions.[12] Such overlaps mean that federal agencies are spending resources on similar activities, which may waste scarce resources and limit effectiveness.

Specifically, we found that FDA food safety activities may overlap with, if not duplicate, the efforts of other agencies, including USDA and NMFS. FDA could take practical steps to reduce overlap and duplication and thereby free resources for more effective oversight of food safety, but FDA has made little progress since our report. For example:

- Domestic inspections. In fiscal year 2003, FDA and USDA spent most of their food safety resources—about $900 million—on inspection and enforcement activities. A portion of these activities included overlapping and even duplicative inspections of 1,451 domestic food-processing facilities that produce foods regulated by both agencies. Under authority granted by the Bioterrorism Act of 2002,[13] FDA could authorize USDA to inspect these facilities on its behalf, but FDA has not yet reached an agreement with USDA to do this. We recommended that, if cost effective, FDA enter into an agreement to commission USDA inspectors at jointly regulated facilities. FDA told us that they are working with USDA to consider which products might be covered by each agency under such an agreement.
- Import inspections. FDA and USDA both inspect shipments of imported food at ports of entry and also visit foreign countries that export food to the United States. We found that both FDA and USDA maintain inspectors at 18 U.S. ports of entry to inspect imported food. In fiscal year 2003, FDA spent more than $115 million on imported food inspections, and USDA spent almost $16 million. The two agencies do not share inspection resources at these ports. Although USDA maintains a daily presence at these facilities, the FDA-regulated products may remain at the facilities for some time awaiting FDA inspection. Further, FDA conducted inspections in 6 of the 34 countries that USDA evaluated in 2004 to determine whether their food safety systems for ensuring the safety of meat and poultry are equivalent to that of the United States. We recommended that FDA consider the findings of USDA's foreign country equivalence agreements when determining which countries to visit. In their response to our recommendation, the agency noted that they will consider USDA's foreign country evaluations when making such determinations.
- Inspectors' training. FDA and USDA spend resources to provide similar training to food inspection personnel. FDA spent about $1.6 million and USDA spent $7.8 million in fiscal year 2003. We found that, to a considerable extent, food inspection training addresses the same subjects, such as plant sanitation and good manufacturing practices. While other

agencies have consolidated training activities that have a common purpose and similar content, FDA and USDA have not. We recommended that USDA and FDA consider joint training programs, but to date, FDA has told us that they have identified no training needs common to both agencies.

FDA's Food Protection Plan Proposes Some Positive First Steps, but Additional Information on the Plan's Strategies and Resources Would Facilitate Congressional Oversight

FDA's Food Protection Plan proposes several positive first steps that are intended to enhance food safety oversight, including requesting several authorities recommended by GAO, but more specific information about its strategies and the resources needed to implement the plan would facilitate congressional oversight. Positively, FDA's Food Protection Plan aims to shift the agency's focus to prevention of foodborne illness instead of intervention after contamination and resulting illnesses occur—an important shift given that experts consider prevention to be a core element of an effective food safety system. FDA says that its key prevention steps are promoting corporate responsibility, identifying food vulnerabilities, assessing risks, and expanding its understanding and use of effective mitigation measures.

In addition to the actions we discuss earlier to address resource constraints, FDA's Food Protection Plan requests other authorities to enhance oversight of food safety that begin to respond to prior GAO recommendations. Specifically, the plan requests authority for FDA to:

- Order food recalls. The Food Protection Plan requests the authority to order a recall when FDA has reason to believe that food is adulterated and presents a threat of serious adverse health consequences or death, to be imposed only if a company refuses or unduly delays conducting a voluntary recall. Currently, food recalls are largely voluntary—federal agencies responsible for food safety, including FDA, have no authority to compel companies to recall contaminated foods, with the exception of FDA's authority to require a recall for infant formula. FDA does have authority, through the courts, to seize, condemn, and destroy adulterated or misbranded food under its jurisdiction and to disseminate information about foods that are believed to present a danger to public health. However, government agencies that regulate the safety of other products, such as toys and automobile tires, have recall authority not available to FDA for food and have had to use their authority to ensure that recalls were conducted when companies did not cooperate. These agencies have the authority to require a company to notify the agency when the company has distributed a potentially unsafe product, order a recall, establish recall requirements, and impose monetary penalties if a company does not cooperate. In a report and testimony before this subcommittee,[14] we noted that limitations in the FDA's food recall authorities heighten the risk that unsafe food will remain in the food supply and have proposed that Congress consider giving FDA similar authorities. While FDA's Food Protection Plan requests mandatory recall authority, this request could also include recall authorities held by other agencies, including establishing recall requirements and imposing penalties for noncompliance. FDA officials noted that while recall requirements and penalties for noncompliance were not explicitly stated in the Food Protection Plan, they are encompassed in the request. Further, the plan does not propose a definition of "undue delay" by a company, another critical element of recall authority given that timing is essential in reacting to outbreaks, and delays can cost lives.

- Issue additional preventive controls for high-risk foods. FDA is requesting explicit authority from Congress to issue regulations requiring foods that have been associated with repeated instances of serious health problems or death to be prepared, packed, and held under a system of preventive food safety controls. According to FDA, this would clarify the agency's ability to require industries to implement preventive Hazard Analysis and Critical Control Point (HACCP) systems, which it currently requires for companies that process seafood and juice. HACCP systems are designed to improve food safety by having industry identify and control hazards in products before they enter the market. FDA officials told us that they are asking for explicit authority to put measures in place for other high-risk foods, such as leafy greens. Officials told us that this request, if granted, would allow the agency to focus its preventive efforts on foods that present the highest risk for contamination, consistent with the agency's risk-based focus. However, others have expressed concern that requiring a history of repeated outbreaks before issuing preventive controls would not allow FDA to proactively establish regulations for foods before they cause additional illnesses.

While FDA officials have acknowledged that implementing the Food Protection Plan will require additional resources, FDA has not provided specific information on the resources it anticipates the agency will need to implement this plan. For example, the Food Protection Plan proposes to develop food protection guidelines for industry; however FDA's Science Board reported that modernizing safety standards for fresh produce and other raw foods and developing and implementing inspection programs could cost $210 million. Additionally, the Food Protection Plan proposes to enhance FDA's information technology systems related to both domestic and imported foods which the Science Board report suggests could cost hundreds of millions of dollars. FDA officials have declined to provide specific information on how much additional funding it believes will be necessary to implement the Food Protection Plan, saying that finalizing the amounts will take place during the budget process. Similarly, the Food Protection Plan does not discuss the strategies it needs in the upcoming years to implement this plan. FDA officials told us that they have internal plans for implementing the Food Protection Plan that detail timelines, staff actions, and specific deliverables. While FDA officials told us they do not intend to make these plans public, they do plan to keep the public informed of their progress. Without a clear description of resources and strategies, it will be difficult for Congress to assess the likelihood of the plan's success in achieving its intended results.

Tools that Agencies can Use to Address Management Challenges

The Science Board cites numerous management challenges that have contributed to FDA's inability to fulfill its mission, such as a lack of a coherent structure and vision, insufficient capacity in risk assessment, and inadequate human capital recruitment and retention. The Science Board also noted that public confidence in FDA's abilities has diminished. In light of these challenges, we have identified through other work some tools that can help agencies improve their performance, which may also be relevant to FDA.

For example, we reported on the use of a Chief Operating Officer (COO)/Chief Management Officer (CMO) as one way to address longstanding management problems that are undermining agencies' abilities to accomplish their missions and achieve results.[15] Agencies with such challenges, including FDA, could benefit from a senior leader serving as a COO/CMO who can elevate, integrate, and institutionalize responsibility for key management functions. While

GAO has long advocated the need for a COO/CMO position at the Department of Defense and the Department of Homeland Security, a relatively stable or small organization could use the existing deputy or related position to carry out the role. In addition to GAO, a number of other organizations have supported the need for the creation of COO/CMO positions in federal agencies. McKinsey and Company recommended that a COO be established in many federal agencies as the means to help those agencies successfully achieve transformation.[16] In addition, a working group within the National Academy of Public Administration (NAPA) recommended creating COO positions in federal agencies to oversee the full range of management functions, including procurement, finance, information technology, and human capital.[17]

Another tool that can help federal agencies address their management challenges is a well-designed commission that can produce specific practical recommendations that Congress can enact. For example, Congress created the National Commission on Restructuring the Internal Revenue Service (IRS) in 1995 to review current practices at IRS and report on requirements for improvement. Congress subsequently passed the IRS Restructuring and Reform Act of 1998, which was influenced by the Commission's report, and reorganized the structure and management of IRS, revised the mission of IRS, and mandated numerous other detailed changes.[18] Based on our recent analysis of several commissions, there are several critical success factors that can be applied to ensure a commission's success including:[19]

- A statutory basis with adequate authority. When provided with a clear mandate and adequate authority, a commission can comprehensively access and analyze information related to a given policy issue and thereby provide more informed policy options for the President and Congress to consider.
- A clear purpose and timeframe. A commission should have a clear purpose for its objectives and activities to help guide the members in carrying out their responsibilities. In addition, a fixed agenda and timeframe can help keep a commission focused and on track. However, a commission should have a broad enough scope to help ensure it has the authority to address all the issues necessary in order to come up with a comprehensive and integrated solution without encountering any constraints in the process as to what it can or cannot consider.
- Key leadership support. Institutional leadership, commitment, and support from the President and Congress are necessary to help a commission succeed.

- An open and transparent process. By having an open and transparent process, such as public hearings, a commission can help build consensus among the public for its goals by gaining their input and support.
- A balanced and capable membership. Balanced and capable membership can help lessen political influences and build consensus among the commission members when carrying out its purpose. Specifically, a commission should involve current or former Members of Congress as well as experts and professionals on the topic. Current or former elected officials can ensure viability of a commission's legislative proposals due to their experience.
- Accountability. Clear accountability for a commission can help foster specific, useful outputs that could help inform the public and provide specific policy options and, hopefully, recommendations for Congress and the President.
- Resources. The success of the commission is dependent on having the adequate resources to carry out its purpose and any potential recommendations.

Generally, one concern regarding commissions may be whether or not there is sufficient buy-in from key stakeholders on the purpose of the commission along with a commitment to act on any resulting recommendations. Any recommendations by a commission in a final report are generally advisory in nature and may not automatically result in any public policy changes. Congressional action through subsequent legislation with Presidential support may be necessary for the commission's recommendations to be implemented and for any changes to occur.

Food safety concerns not only continue but will likely become more urgent in view of changing demographics and consumption patterns. Clearly, FDA plays a critical role in the federal oversight of food safety because of the breadth of its responsibilities. Thus its ability to carry out those responsibilities is necessary to help ensure the safety of the nation's food supply in the most efficient, effective, accountable, and sustainable way. Nevertheless, in light of the federal government's long-term fiscal challenges, agencies, including FDA, need to seek out opportunities to better leverage their resources. FDA's Food Protection Plan is a step in the right direction and proposes to implement many of the recommendations made by GAO. However, additional information on the strategies and resources needed to implement the plan can help Congress assess the likelihood of its success. Further, concerns over FDA's management challenges, such as those identified by the Science Board could hinder the implementation of the plan. Tools such as commissions and positions like a COO/CMO can help agencies address management challenges and make needed progress to achieve their missions. Continued congressional oversight, including today's hearing, and additional legislative action are key to achieving that progress and to promoting the safety and integrity of the nation's food supply.

Mr. Chairman, this concludes my prepared statement. I would be pleased to respond to any questions that you or other Members of the Subcommittee may have.

Contact and Staff Acknowledgments

Contact points for our Offices of Congressional Relations and Public Affairs may be found on the last page of this statement. For further information about this testimony, please contact Lisa Shames, Director, Natural Resources and Environment at (202) 512-3841 or shamesl@gao.gov. Key contributors to this statement were Candace Carpenter, Bart Fischer, José Alfredo Gómez, and Alison O'Neill.

References

[1] To view the full product, including the scope and methodology, click on GAO-08-435T. For more information, contact Lisa Shames at (202) 512-3841 or ShamesL@gao.gov.

[2] Department of Health and Human Services, U.S. Food and Drug Administration, *Food Protection Plan* (Washington, D.C., 2007).

[3] Interagency Working Group on Import Safety, Action Plan for Import Safety (Washington, D.C., 2007).

[4] [3]FDA Science Board, Subcommittee on Science and Technology, *FDA Science and Mission at Risk* (Washington, D.C., November 2007).

[5] GAO, Medical Devices: Challenges for FDA in Conducting Manufacturer Inspections, GAO-08-428T (Washington, D.C.: Jan. 29, 2008).

[6] GAO, High-Risk Series: An Update, GAO-07-310 (Washington, D.C.: Jan. 31, 2007).

[7] GAO, Overseeing the U.S. Food Supply: Steps Should be Taken to Reduce Overlapping Inspections and Related Activities, GAO-05-549T (Washington, D.C.: May 17, 2005).

[8] Institute of Medicine, Ensuring Safe Food from Production to Consumption (Washington, D.C., 1998).

[9] GAO, Food Safety and Security: Fundamental Changes Needed to Ensure Safe Food, GAO-02-47T (Washington, D.C.: Oct. 10, 2001).

[10] According to FDA, an entry line is each portion of an import shipment that is listed as a separate item on an entry document. Items in an import entry having different tariff descriptions must be listed separately.

[11] GAO, Food Safety: FDA's Imported Seafood Safety Program Shows Some Progress, but Further Improvements Are Needed, GAO-04-246 (Washington, D.C.: Jan. 30, 2004).

[12] GAO, Oversight of Food Safety Activities: Federal Agencies Should Pursue Opportunities to Reduce Overlap and Better Leverage Resources, GAO-05-213 (Washington, D.C.: Mar. 30, 2005).

[13] Public Health Security and Bioterrorism Preparedness and Response Act of 2002, Pub. L. No. 107-188, 116 Stat. 594.

[14] GAO, Food Safety: USDA and FDA Need to Better Ensure Prompt and Complete Recalls of Potentially Unsafe Food, GAO-05-51 (Washington, D.C.: Oct. 6, 2004) and Federal Oversight of Food Safety: High Risk Designation Can Bring Attention to Limitations in the Federal Government's Food Recall Programs, GAO-07-785T (Washington, D.C.: Apr. 24, 2007).

[15] See for example, GAO, A Call for Stewardship: Enhancing the Federal Government's Ability to Address Key Fiscal and Other 21st Century Challenges, GAO-08-93SP (Washington D.C.: November 2007) and Organizational Transformation: Implementing Chief Operating Officer/Chief Management Officer Positions in Federal Agencies, GAO-08-34 (Washington, D.C.: Nov. 1, 2007).

[16] McKinsey and Company, How Can American Government Meet its Productivity Challenge? (July 2006).

[17] NAPA, Moving from Scorecard to Strategic Partner: Improving Financial Management in the Federal Government (October 2006).

[18] Pub. L. No. 105-206 (July 22, 1998).

[19] GAO, Long-Term Fiscal Challenge: Comments on the Bipartisan Task Force for Responsible Fiscal Action Act, GAO-08-238T (Washington, D.C.: Oct. 31, 2007).

In: Food Protection and Safety
Editor: Lyman F. Thomasson
ISBN 978-1-60692-187-6

Chapter 3

FEDERAL OVERSIGHT OF FOOD SAFETY. FDA HAS PROVIDED FEW DETAILS ON THE RESOURCES AND STRATEGIES NEEDED TO IMPLEMENT ITS FOOD PROTECTION PLAN*

Lisa Shames

WHY GAO DID THIS STUDY[1]

The Food and Drug Administration (FDA) is responsible for ensuring the safety of roughly 80 percent of the U.S. food supply, including $417 billion worth of domestic food and $49 billion in imported food annually. Changing demographics and consumption patterns along with an increase in imports have presented challenges to FDA. At the same time, recent outbreaks, such as E. coli from spinach and Salmonella from tomatoes, have undermined consumer confidence in the safety of the food supply. In November 2007, FDA released its Food Protection Plan, which articulates a framework for improving food safety oversight. In January 2008, GAO expressed concerns about FDA's capacity to implement the Food Protection Plan and noted that more specific information about the strategies and resources needed to implement the plan would facilitate congressional oversight.

* Excerpted from GAO Report-08-909T, dated June 12, 2008.

This testimony focuses on (1) FDA's progress in implementing the Food Protection Plan, (2) FDA's proposal to focus inspections based on risk, and (3) FDA's implementation of previously issued GAO recommendations intended to improve food safety oversight. To address these issues, GAO reviewed FDA documents, such as FDA's operations plan, and FDA data related to the plan. GAO also interviewed FDA officials regarding the progress made. GAO also analyzed FDA data on domestic and foreign food firm inspections. GAO also analyzed the status of past recommendations.

What GAO Found

Since FDA's Food Protection Plan was first released in November 2007, FDA has added few details on the resources and strategies required to implement the plan. FDA plans to spend about $90 million over fiscal years 2008 and 2009 to implement several key actions, such as identifying food vulnerabilities and risk. From the information GAO has obtained on the Food Protection Plan, however, it is unclear what FDA's overall resource need is for implementing the plan, which could be significant. For example, based on FDA estimates, if FDA were to inspect each of the approximately 65,500 domestic food firms regulated by FDA once, the total cost would be approximately $524 million. In addition, timelines for implementing the various strategies in the plan are also unclear, although a senior level FDA official estimated that the overall plan will take 5 years to complete. Importantly, GAO has noted that public reporting is the means through which the federal government communicates the results of its work to the Congress and the American people. FDA officials told GAO that they had prepared a draft report on progress made in implementing the Food Protection Plan, but as of June 4, 2008, FDA told GAO that the Department of Health and Human Services had not cleared the report for release.

The Food Protection Plan identifies the need to focus safety inspections based on risk, which is particularly important as the numbers of food firms have increased while inspections have decreased. For example, between 2001 and 2007, the number of domestic firms under FDA's jurisdiction increased from about 51,000 to more than 65,500, while the number of firms inspected declined slightly, from 14,721 to 14,566. Thus, conducting safety inspections based on risk has the potential to be an efficient and effective approach for FDA to target scarce resources based on relative vulnerability and risk.

FDA has implemented few of GAO's past recommendations to leverage its resources and improve food safety oversight. Since 2004, GAO has made a total of 34 food safety related recommendations to FDA, and as of May 2008, FDA has implemented 7 of these recommendations. For the remaining recommendations, FDA has not fully implemented them, however, in some cases, FDA has taken some steps. However, the planned activities in the Food Protection Plan could help address several of the recommendations that FDA has not implemented. For example, in January 2004, GAO recommended that FDA make it a priority to establish equivalence agreements with other countries. We found that such agreements would shift some of FDA's oversight burden to foreign governments. As of May 2008, FDA has not yet established equivalence agreements with any foreign countries. The Food Protection Plan requests that Congress allow the agency to enter into agreements with exporting countries to certify that foreign producers' shipments of designated high-risk products comply with FDA standards.

Mr. Chairman and Members of the Subcommittee:

I am pleased to be here today to discuss the Food and Drug.Administration's (FDA) progress in implementing the Food Protection Plan, which articulates FDA's plans to improve the oversight of food safety. FDA is responsible for ensuring the safety of roughly 80 percent of the U.S. food supply—virtually all foods except for meat, poultry, and processed egg products—including $417 billion worth of domestic food and $49 billion in imported food annually. As you know, in January 2007, we designated the federal oversight of food safety as a high-risk area needing urgent attention and transformation.[2] A key reason for that designation is that FDA is one of 15 agencies that collectively administer at least 30 laws related to food safety. Around the time of this designation, consumers faced several outbreaks of foodborne illnesses, including E. coli from spinach and Salmonella from peanut butter. Subsequently, the U.S. has seen more outbreaks of foodborne illnesses, such as Salmonella from imported cantaloupes and raw tomatoes. Not surprisingly, public trust in FDA's ability to protect the food supply has fallen. A 2008 Harris poll showed that U.S. adults have little confidence—and less confidence than last year—in the safety of packaged or prepared foods that have been imported from countries like China, India, or South Africa. In addition, a recent public opinion poll conducted by the Trust for America's Health[3] found that 67 percent of Americans are worried about food safety, ranking it higher than concerns about, for example, pandemic flu or natural disasters.

Concerns about food safety oversight are not new. GAO and others have consistently reported on a lack of adequate oversight of food safety by FDA, and have provided many recommendations for better leveraging FDA's limited resources and suggestions for additional authorities that would allow FDA to better fulfill its responsibilities. In 1998, we reported that limitations in FDA's authority and its need to more effectively target limited resources could adversely affect its ability to ensure food safety.[4] A decade later, the story remains the same and has only taken on a greater sense of urgency due to changing demographics and consumption patterns that, according to FDA, have put more of the U.S. population at risk of contracting foodborne illnesses. Populations at high risk of foodborne illnesses—older adults, young children, pregnant women, and immune compromised individuals—now make up 20 to 25 percent of the U.S.

population. In addition, U.S. consumers are increasingly eating raw or minimally processed foods, which are often associated with foodborne illnesses. For example, the consumption rate of leafy greens—the category of produce most likely to be associated with an outbreak—increased 180 percent between 1992 and 2005, according to the U.S. Department of Agriculture. Compounding the challenges, the number of FDA-regulated domestic food establishments has increased more than 10 percent in the last five years and the number of food import entry lines has tripled in the past ten years.[5]

To respond to the need for better oversight of food safety, FDA released its Food Protection Plan in November 2007, which articulates FDA's framework for overseeing the safety of food and outlines three core elements—prevention, intervention, and response—that are the focus of FDA's efforts to improve oversight.[6] At the same time, a twelve-agency working group presented to the President its Action Plan for Import Safety,[7] which contains, among other things, recommendations for improving the safety of food imports entering the United States. Both plans spell out numerous actions FDA plans to take to enhance food safety, including writing new food protection guidelines for industry and helping foreign countries improve their regulatory systems. The plans also request new legislative authorities, such as enhanced access to a food company's records during food safety emergencies.

Also, in November 2007, FDA's Science Board, an advisory board to the agency, released a report entitled, FDA Science and Mission at Risk.[8] This report concluded that FDA is not positioned to meet current or emerging regulatory needs, and stated that FDA does not have the capacity, such as staffing and technology, to ensure the safety of the nation's food supply. According to the Science Board report, FDA's resources have not kept pace with its increasing responsibilities, and this disparity has made it increasingly "impossible" for FDA to maintain its historic public health mission. In addition, the report finds that food safety resources have increasingly been diverted away from routine surveillance and other tasks to managing crises as they arise and the nation's food supply is at risk. In February 2008, the Science Board estimated that, to implement its recommendations to protect the nation's food supply, FDA's base budget would need to increase by a total of $755 million by fiscal year 2013, phased in over time starting with $128 million in fiscal year 2009.

In response to these concerns, Congress has expressed considerable interest in enhancing FDA's oversight of food safety, and the House Energy and Commerce Committee has held hearings to consider a draft bill entitled The Food and Drug Administration Globalization Act of 2008 which, in part, would provide some of FDA's requested authorities.[9] This draft bill also contains provisions that are consistent with several past GAO recommendations to FDA and matters for congressional consideration regarding FDA's food safety programs. For example, the draft bill contains provisions that would allow FDA to leverage resources using outside organizations, such as third-party inspectors.

As part of its congressional oversight of FDA's challenges in meeting its responsibilities, we testified in January 2008 before this subcommittee and reported that FDA's Food Protection Plan proposes positive first steps for FDA.[10] For example, FDA requests authority to order food recalls and issue additional preventive controls for high-risk foods, both of which we previously recommended. However, we expressed concerns about FDA's capacity to implement the plan and noted that more specific information about its strategies and the resources FDA needs to implement the plan would facilitate congressional oversight. We recognized that without a clear description of resources and strategies, it would be difficult for Congress to assess the likelihood that the plan will achieve its intended results.

In this context, my testimony today focuses on FDA's progress in implementing the Food Protection Plan, FDA's proposal to focus inspections based on risk, and FDA's implementation of previously issued GAO recommendations intended to improve food safety oversight. In summary, we have found (1) FDA has added few details on the resources and strategies required to implement its Food Protection Plan, (2) FDA's proposal to focus inspections based on risk can help target scarce resources, and (3) FDA has implemented few of our recommendations intended to help leverage resources and improve operations. This testimony is based on new and previously issued work.

To assess FDA's progress in implementing the Food Protection Plan, we reviewed FDA documents, such as FDA's operations plan and work plan, and FDA data related to the plan. In addition, we interviewed FDA officials regarding the progress made to date in implementing the Food Protection Plan. To review FDA's proposal to focus inspections based on risk, we analyzed FDA's data on past domestic and foreign food firm inspections. To determine actions that FDA has taken on our past recommendations, we obtained and analyzed information from FDA on the status of these recommendations. We conducted our work between May and June 2008 in accordance with generally accepted government auditing standards. Those standards require that we plan and perform the audit to obtain sufficient, appropriate evidence to provide a reasonable basis for our findings and conclusions based on our audit objectives. We believe the evidence obtained provides a reasonable basis for our findings and conclusions based on our audit objectives.

FDA HAS ADDED FEW DETAILS ON THE RESOURCES AND STRATEGIES REQUIRED TO IMPLEMENT ITS *FOOD PROTECTION PLAN*

In light of the federal government's long-term fiscal challenges, it is critical that agencies can justify the needed resources and develop effective, efficient strategies to achieve their mission. We testified in January 2008 that, while FDA officials had acknowledged that implementing the Food Protection Plan would require additional resources, FDA had not provided specific information on the resources it anticipates the agency will need to implement this plan to improve its oversight of food safety. For example, the Food Protection Plan proposes to enhance FDA's information technology systems related to both domestic and imported foods which the Science Board report suggests could cost hundreds of millions of dollars. At that time, FDA officials stated they would provide specific information on how much additional funding would be necessary to implement the Food Protection Plan when the President's budget was publicly released in the coming weeks.

In its fiscal year 2008 budget, FDA received approximately $620 million for food protection, an increase of about $56 million over fiscal year 2007, and directed $48 million of that amount toward implementing the Food Protection Plan, according to FDA. FDA requested approximately $662 million for food safety for fiscal year 2009, an increase of about $42 million over fiscal year 2008. According to the Department of Health and Human Services' budget justification, FDA plans to direct the $42 million to strategic actions described in its Food Protection Plan. As shown in table 1, the plan outlines spending on all three core elements of the Food Protection Plan—a total of about $21 million for prevention, about $34 million for intervention, and about $23 million for response for fiscal years 2008 and 2009. FDA also reported that, in fiscal year 2008, the agency intends to hire nearly 1,500 full time equivalents (FTE), including approximately 730 to fill vacant positions. Of these, 161 will be new FTEs funded by congressional increases dedicated to food safety activities. In addition, in fiscal year 2009, FDA plans to hire 94 new FTEs for food safety activities.

Furthermore, in May 2008, FDA's Commissioner of Food and Drugs provided his professional judgment in response to a congressional request of FDA's immediate resource needs to implement key initiatives across the core elements of the Food Protection Plan. The Commissioner called for an additional $125 million for food protection in fiscal year 2008 beyond the $48 million that FDA had already allocated for implementing the Food Protection Plan in this fiscal year. According to the Commissioner, this increase will allow FDA to address some of the plan's strategic actions, such as identifying and targeting the greatest threats from intentional and unintentional contamination and conducting more risk-based inspections. The Commissioner's assessment also calls for 250 additional FTEs to accomplish the goals of the Food Protection Plan. After the Commissioner provided his assessment of FDA's resource needs, the Senate passed an Iraq War Supplemental that included an additional $119 million for food safety to be available through fiscal year 2009. In addition, on June 9, 2008, the Department of Health and Human Services announced that the Administration is amending its fiscal year 2009 budget request to include, in part, a $125 million increase for food safety. This amount would add to the $42 million increase originally proposed in the fiscal year 2009 budget justification (see table 1) and appears to be consistent with the

Commissioner's professional judgment response. To accompany this amendment, FDA has posted information on steps it is taking to invest in its transformation in areas such as domestic medical products, import products, and domestic food safety. For example, under transforming domestic food safety, FDA reports that it issued final fresh cut produce guidance to limit contamination of fresh-cut fruits and vegetables. In addition, FDA conducted inspections and took action against processors of low acid canned foods that were deviating from required standards.

Table 1. Current and planned spending for *Food Protection Plan* Core Elements and Strategic Actions, Fiscal Years 2008 and 2009

Food Protection Plan core elements and strategic actions	Fiscal year 2008 increase	Fiscal year 2009 increase	Total current/planned spending for fiscal years 2008 and 2009
Total for core element 1: prevention	10,024,000	11,414,000	21,438,000
1.1: Promote increased corporate responsibility to prevent foodborne illnesses	3,108,000	6,311,000	9,419,000
1.2: Identify food vulnerabilities and assess risks	5,580,000	4,302,000	9,882,000
1.3: Expand the understanding and use of effective mitigation measures	1,336,000	801,000	2,137,000
Total for core element 2: intervention	18,509,000	15,606,000	34,115,000
2.1: Focus inspections and sampling based on risk	16,187,000	14,864,000	31,051,000
2.2: Enhance risk-based surveillance of imported foods at the border	2,322,000	742,000	3,064,000
2.3: Improve the detection of food system "signals" that indicate contamination	0	0	0
Total for core element 3: response	19,589,000	3,174,000	22,763,000
3.1: Improve immediate response	19,589,000	2,954,000	22,543,000
3.2: Improve risk communications to the public, industry and other stakeholders	0	220,000	220,000
Sub-total	48,122,000	30,194,000	78,316,000
Cost of living pay increase for onboard food protection employees	0	12,038,000	12,038,000
Total for entire *Food Protection Plan*	48,122,000	42,232,000	90,354,000

Source: FDA.

In addition, in January 2008, we testified that the Food Protection Plan does not discuss the strategies it needs in the upcoming years to implement this plan. When we asked FDA for more specificity on the strategies for implementing the plan, FDA officials told us that they have internal plans for implementing the Food Protection Plan that detail timelines, staff actions, and specific deliverables. More recently, a senior level FDA official provided us with an estimate of 5 years for fully implementing the plan. However, FDA has not provided us with timelines for the various strategies described in the plan. For example, under the plan's strategic action 2.3—to improve the detection of food system "signals" that indicate contamination (see table 1)—FDA has recently identified three additional action steps with deliverables that will be needed to identify, develop, and deploy new screening tools and methods to identify pathogens and other contaminants. However, FDA could not provide us with an estimate of how long it would take to implement these steps or the overall strategic action. Without this type of information, we are not able to assess whether FDA's estimated 5-year time frame is feasible.

Similarly, while FDA's Food Protection Plan recognizes the need to partner with Congress to obtain 10 additional statutory authorities to transform the safety of the nation's food supply, FDA's congressional outreach strategy is general. When we asked FDA officials if they had a congressional outreach strategy, FDA officials told us that they had met with various congressional committees to discuss the Food Protection Plan. When asked if they had provided draft language to congressional committees on the various authorities, FDA officials explained that they only provided technical assistance, such as commenting on draft bills, to congressional staff when asked FDA appears to be refining its implementation plan over time. Most recently, in June 2008, FDA provided us with a draft work plan that it characterizes as a dynamic document that changes on a daily basis to implement the Food Protection Plan. While this draft work plan provides more information on the action steps and deliverables to achieve the core elements, we continue to have concerns about FDA's lack of specificity on the necessary resources and strategies to fully implement the plan. For example, as part of the plan's strategic action 1.1—to promote increased corporate responsibility to prevent foodborne illnesses (see table 1)—FDA has identified a goal of analyzing food import trend data and focusing inspections based on risk, and the draft work plan shows six deliverables, such as analysis of import data sets and an import risk ranking, associated with this goal. However, the timelines for these deliverables are unclear. In addition, the agency plans to dedicate a total of $673,000 to this goal in fiscal years 2008 and 2009, and FDA officials told us that the agency considers this funding to be a down payment toward achieving this goal. However, it is unclear what the total cost will be to meet this goal. While the work plan provides some basic information, more specific information, such as estimated resources needed to implement the various strategies— the core elements, goals, and deliverables—as well as the overall plan and timeframes for implementing the strategies, are needed to assess FDA's progress in implementing the plan or in acquiring the resources and authorities it needs.

Anticipating the cost of the overall plan is important because, while some activities, such as meeting with industry experts to discuss corporate responsibility, may be accomplished within one budget cycle, others, such as the establishment of an FDA field office in China will likely require a long-term commitment of agency resources. From the information we have obtained on the Food Protection Plan, it is unclear what FDA's overall resource need is for implementing the plan. The overall resource need could be significant. For example, if FDA were to inspect each of the approximately 65,500 domestic food firms regulated by FDA, at the Commissioner's May 2008 estimate of $8,000 for a domestic food safety inspection, it would cost approximately $524 million to inspect all of these facilities once. Similarly, if FDA were to inspect each of the 189,000 registered foreign facilities (which includes facilities that manufacture, process, pack, or hold foods consumed by Americans) at the

Commissioner's estimated cost of $16,700 per inspection, it would cost FDA approximately $3.16 billion to inspect all of these facilities once. These figures underscore the need for FDA to focus safety inspections based on risk.

Ultimately, a results-oriented organization needs to take a long-term view of the goals it wants to accomplish and describe them in a strategic plan. To facilitate congressional oversight, strategic plans should discuss (1) long-term goals and objectives for all major functions; (2) approaches to achieve the goals and objectives, and in particular the required resources including human capital and information technology; (3) a relationship between the long-term goals and the annual performance goals; and (4) an identification of key factors that could significantly affect achievement of the strategic goals. Such discussions in the Food Protection Plan could help clarify FDA's organizational priorities to the Congress, other stakeholders, and the public.

Lastly, when we testified before this subcommittee in January, we reported that FDA planned to keep the public informed of their progress on implementing the Food Protection Plan. In addition, in March 2008, FDA officials indicated that a progress report on actions taken to implement the Food Protection Plan would be issued in April 2008. In May, FDA officials told us that they had prepared a draft progress report, but as of June 4, 2008, FDA had not made this report public. FDA officials told us that the progress report is still being cleared by the

Department of Health and Human Services, and they could not provide us with the report until it was cleared by the department. Instead, FDA officials provided us with a broad overview of FDA's actions and, subsequently, provided us with a list of accomplishments drawn out of numerous public documents. For example, FDA issued a Federal Register Notice to solicit stakeholder comments on the implementation of the Food Protection Plan as part of a broad outreach plan.

We have noted that public reporting is the means through which the federal government communicates the results of its work to the Congress and the American people. Such reporting is in the public interest and promotes transparency in government operations. While it is important to show what progress has been made, having such information in a consolidated document at a readily accessible location reassures Congress and the public that actions have been taken.

FDA's Proposal to Focus Inspections Based on Risk can Help Target Scarce Resources

The Food Protection Plan identifies the need to focus safety inspections based on risk, which is particularly important as the numbers of food firms have increased while inspections have decreased. In its Food Protection Plan, FDA has identified some actions to better identify food vulnerabilities and assess risks. For example, FDA plans to use enhanced modeling capability, scientific data, and technical expertise to evaluate and prioritize the relative risks of specific food and animal feed agents that may be harmful. According to FDA officials, the agency has assigned a risk-based steering committee to identify models for ranking and prioritizing risk.

Conducting inspections based on risk has the potential to be an efficient and effective approach for FDA to target scarce resources, particularly when the number of inspections has not kept pace with the growth in firms between 2001 and 2007. Specifically, while the number of domestic firms under FDA's jurisdiction increased from about 51,000 to more than 65,500, the number of firms inspected declined slightly, from 14,721 to 14,566. FDA also reported declines in the number of inspections at overseas firms between 2001 and 2007—even as the United States has imported hundreds of thousands of different food products from tens of thousands of foreign food firms in more than 150 countries. Appendix I has information on the number of FDA inspections of food firms in foreign countries from fiscal years 2001 through 2007.

GAO HAS ISSUED RECOMMENDATIONS INTENDED TO HELP LEVERAGE RESOURCES AND IMPROVE OPERATIONS, BUT FDA HAS IMPLEMENTED FEW OF THEM

FDA has implemented few of our past recommendations to improve food safety oversight. Our recommendations are designed to correct identified problems and improve programs and operations. We have made 34 food safety related recommendations to FDA since 2004 and, as of May 2008, FDA has implemented 7. For the remaining recommendations, FDA has not fully implemented them, however, in some cases, FDA has taken some steps. As shown in table 2, these recommendations fall into two broad categories: improving monitoring and enforcement processes and leveraging resources. The planned activities in the Food Protection Plan could help address several of these recommendations.

Table 2. FDA's Implementation of GAO's Food Safety Recommendations, Since 2004

Category of recommendation	Total recommendations	Recommendations FDA has implemented	Recommendations FDA has not fully implemented
Improving monitoring and enforcement processes	21	3	18
Leveraging resources	13	4	9
Total Recommendations	34	7	27

Source: GAO and FDA.

In light of the federal government's long-term fiscal challenges, agencies, including FDA, need to seek out opportunities to better leverage their resources. We have made 13 recommendations to help FDA better leverage its resources since 2004, and FDA has implemented 4 of them. In a January 2004 report regarding seafood safety, we recommended that, among other things, FDA make it a priority to establish equivalence agreements with other countries.[11] We found that such agreements would shift some of FDA's oversight burden to foreign governments. FDA did not concur with this recommendation, and as of May 2008, has not yet established equivalence agreements with any foreign countries. In the same report, we recommended that FDA give priority to taking enforcement actions when violations that pose the most serious health risk occur; consider the cost and benefits of implementing an accreditation program for private laboratories; and explore the potential of implementing a certification program for third-party inspectors. Although FDA concurred with these recommendations and has taken some limited action such as requesting public comments on the use of third-party certification programs, none were fully implemented. The Food Protection Plan requests that Congress allow the agency to enter into agreements with exporting countries to certify that foreign producers' shipments of high-risk products comply with FDA standards.

Since 2004, we have made 21 recommendations to FDA to improve monitoring and enforcement processes, and FDA has implemented 3 of them. For example, in October 2004, we recommended that FDA develop a sound methodology for district staff to verify that companies have quickly and effectively carried out recalls.[12] At the time of our review, we found that FDA was not calculating the recovery rate for recalls. As a result, the agency did not know how much food was actually recovered, although the agency told us recovery was an important indicator of a successful recall. FDA initially commented that we had not demonstrated that weaknesses in FDA's recall process resulted in little recovery of food, but as of May 2008, the agency is in the process of conducting a quality management system review of its recall activities and, once the review is completed, it will include recommendations for verifying that a company's recall was effective, according to FDA.

To conclude, FDA's release of the Food Protection Plan is a positive first step toward modernizing FDA's approach to food safety to better meet the challenges of an increasingly global food supply and respond to shifting demographics and consumption patterns. Given that FDA's resources have not kept pace with its increasing responsibilities, FDA's plan to take a risk- based approach to inspections could help FDA make the most effective and efficient use of its limited resources. However, FDA's Food Protection Plan can only be as effective as its implementation, and without specificity on the resources and strategies needed to fully implement the plan—and in the absence of public reporting—neither Congress nor the public can gauge the plan's progress or assess its likelihood of success in achieving its intended results. In addition, no one is better poised than FDA to identify the resources and authorities needed to implement the plan; therefore, FDA's capacity to provide such information can be questioned. Meanwhile, as foodborne illness outbreaks continue, FDA is missing valuable opportunities to reassure Congress and the public that it is doing all it can to protect the nation's food supply.

Mr. Chairman, this concludes my prepared statement. I would be pleased to respond to any questions that you or other Members of the Subcommittee may have.

Appendix I. Number of FDA Inspections of Food Firms in Foreign Countries, as of December 2007

FDA Inspections of Foreign Food Firms, Fiscal Years 2001 – 2007

	2001	2002	2003	2004	2005	2006	2007	Total
Mexico	17	15	8	15	7	16	26	104
Ecuador	8		11	24		11	10	64
Chile	13		15	6	7	11		52
Peru	13			18	1	9	9	50
Brazil		12	6	7	21			46
Thailand	4	10		10		22		46
Canada	13		13	1		7	4	38
China		9	2	6	16			33
Taiwan	9	7		9		7		32
Argentina	7	5					19	31
India	6		10		7	7		30
South Korea	14			1	7		6	28
Australia	12		6			9		27
Costa Rica		11		4	5	7		27
Vietnam		9		10	8			27
Honduras	9	8			7			24

FDA Inspections of Foreign Food Firms, Fiscal Years 2001 – 2007 (Continued)

	2001	2002	2003	2004	2005	2006	2007	Total
Fiji			8				13	21
Singapore	10			8				18
Estonia	8			8				16
Guatemala		10			6			16
South Africa	5		11					16
Germany	5	4	4			1	1	15
Nicaragua		8				7		15
31 additional countries[a]	58	61	54	26	40	11	8	258
Total number of countries	26	22	22	20	16	15	11	54
Total Inspections	211	169	148	153	132	125	96	1034

Source: GAO analysis of FDA data.

Note: [a]Countries with a total of 14 or fewer inspections between 2001 and 2007 are not listed in the table. These countries include: El Salvador (14 inspections), Jamaica (14), Latvia (14), Uruguay (14), Venezuela (14), Italy (13), Morocco (13), New Zealand (13), Poland (13), Norway (11), France (10), Romania (10), Suriname (10), Iceland (9), Malaysia (9), Bulgaria (8), Columbia (8), Cyprus (7), Panama (7), Trinidad and Tobago (7), United Kingdom (6), Turkey (5), Spain (4), Belgium (3), Greece (3), Hungary (3), Finland (2), Haiti (2), Japan (2), and the Netherlands (2). FDA also inspected food firms in Hong Kong (8).

REFERENCES

[1] To view the full product, including the scope and methodology, click on GAO-08-909T. For more information, contact Lisa Shames at (202) 512-3841 or shamesl@gao.gov.

[2] GAO, High-Risk Series: An Update, GAO-07-310 (Washington, D.C.: January 2007).

[3] Trust for America's Health is a non-profit, non-partisan organization dedicated to protecting the public's health.

[4] GAO, Food Safety: Federal Efforts to Ensure the Safety of Imported Foods Are Inconsistent and Unreliable, GAO/RCED-98-103 (Washington, D.C.: April 30, 1998)

[5] According to FDA, an entry line is each portion of an import shipment that is listed as a separate item on an entry document. Items in an import entry having different tariff descriptions must be listed separately.

[6] Department of Health and Human Services, U.S. Food and Drug Administration, Food Protection Plan (Washington, D.C., 2007).

[7] Interagency Working Group on Import Safety, Action Plan for Import Safety (Washington, D.C., 2007).

[8] FDA Science Board, Subcommittee on Science and Technology, *FDA Science and Mission at Risk* (Washington, D.C., November 2007).

[9] Committee on Energy and Commerce, U.S. House of Representatives, Discussion Draft: The Food and Drug Administration Globalization Act of 2008 (Apr. 16, 2008), available at http://energycommerce.house.gov/FDAGlobalAct-08/index.shtml.

[10] GAO, Federal Oversight of Food Safety: FDA's Food Protection Plan Proposes Positive First Steps, but Capacity to Carry Them Out is Critical, GAO-08-435T (Washington, D.C.: Jan. 29, 2008).

[11] GAO, *Food Safety: FDA's Imported Seafood Safety Program Shows Some Progress, but Further Improvements are Needed,* GAO-04-246 (Washington, D.C.: Jan. 30, 2004).

[12] GAO, *Food Safety: USDA and FDA Need to Better Ensure Prompt and Complete Recalls of Potentially Unsafe Food*, GAO-05-51 (Washington, D.C.: Oct. 6, 2004).

In: Food Protection and Safety
Editor: Lyman F. Thomasson
ISBN 978-1-60692-187-6

Chapter 4

THE FEDERAL FOOD SAFETY SYSTEM: A PRIMER*

Geoffrey S. Becker and Donna V. Porter

ABSTRACT

Numerous federal, state, and local agencies share responsibilities for regulating the safety of the U.S. food supply, which many experts say is among the safest in the world. Nevertheless, critics view this system as lacking the organization and resources to adequately combat foodborne illness, which sickens an estimated 76 million people and kills an estimated 5,000 each year in this country. The 1 10th Congress may face calls for a review of federal food safety agencies and authorities, and proposals for reorganizing them have been introduced (e.g., H.R. 1148; S. 654). Other bills that would alter some agency responsibilities include S. 1274, H.R. 2108, S. 1082, H.R. 1600, H.R. 2629, and S. 887. Among the issues likely to arise are whether reform can improve oversight, and the cost to industry, consumers, and taxpayers. This chapter provides a brief introduction to the system and the debate on whether reorganization is needed.

---

* Excerpted from CRS Report RS22600 dated June 18, 2007.

BACKGROUND [1]

Americans spent more than $1 trillion on food in 2005, nearly half of it in restaurants, schools, and other places outside the home.[2] Federal laws give food manufacturers, distributors, and retailers the basic responsibility for assuring that foods are wholesome, safe, and handled under sanitary conditions. A number of federal agencies, cooperating with state, local, and international entities, play a major role in regulating food quality and safety under these laws.

The combined efforts of the food industry and the regulatory agencies often are credited with making the U.S. food supply among the safest in the world. Nonetheless, public health officials estimate that each year 76 million people become sick, 325,000 are hospitalized, and 5,000 die from foodborne illnesses caused by contamination from any one of a number of microbial pathogens.3 At issue is whether the current system has the resources and structural organization to protect consumers from these dangers. Also at issue is whether federal food safety laws, first enacted in the early 1900s, have kept pace with the significant changes that have occurred in the food production, processing, and marketing sectors since then.

THE AGENCIES AND THEIR ROLES

The Government Accountability Office (GAO) has identified 15 federal agencies collectively administering at least 30 laws related to food safety. The Food and Drug Administration (FDA), which is part of the U.S. Department of Health and Human Services (HHS), and the Food Safety and Inspection Service (FSIS), which is part of the U.S. Department of Agriculture (USDA), together comprise the majority of both the total funding and the total staffing of the government's food regulatory system.[4] FSIS's FY2006 budget was approximately $980 million, $832 million of it appropriated funds and the balance industry-paid user fees. FDA's budget for foods was approximately $440 million in FY2006; oversight of animal drugs and feeds totaled another $99 million, of which approximately $90 million was appropriated and the balance user fees.[5]

Among other agencies with smaller but still significant shares of the food safety portfolio are the National Marine Fisheries Service (NMFS), which is part of the U.S. Department of Commerce (DOC), the Environmental Protection Agency (EPA), and the Centers for Disease Control and Prevention (CDC) in HHS.

Food and Drug Administration

The FDA is responsible for ensuring that all domestic and imported food products — except for most meats and poultry — are safe, nutritious, wholesome, and accurately labeled. Examples of FDA-regulated foods are produce, dairy products, seafood, and processed foods. FDA has jurisdiction over meats from animals or birds that are not under the regulatory jurisdiction of FSIS. FDA shares responsibility for the safety of eggs with FSIS. FDA has jurisdiction over establishments that sell or serve eggs or use them as an ingredient in their products. FDA is also responsible for ensuring that seafood products, including those from aquaculture, do not endanger public health. The primary statutes governing FDA's activities are the Federal Food, Drug, and Cosmetic Act, as amended (21 U.S.C. 301 et seq.); the Public Health Service Act, as amended (42 U.S.C. 201 et seq.); and the Egg Products Inspection Act, as amended (21 U.S.C. 1031 et seq.).

FDA's food inspection force numbers more than 1,900 in field offices throughout the United States, plus nearly 900 in the Washington, DC, area. FDA regulates food manufacturers' safety practices by relying on companies' self-interest in producing safe products, and by working with the industry to improve production practices. About 57,000 food manufacturers are subject to periodic FDA inspection for regulatory compliance. According to GAO, unannounced compliance inspections of individual establishments by FDA officials occur roughly once every five years. FDA relies on notifications from within the industry or from other federal or state inspection personnel, as well as other sources, to alert it to situations calling for increased inspection.

In the Washington, DC, area, two FDA offices are the focal point for food safety- related activities. The Center for Food Safety and Applied Nutrition (CFSAN) is responsible for (1) conducting and supporting food safety research; (2) developing and overseeing enforcement of food safety and quality regulations; (3) coordinating and evaluating FDA's food surveillance and compliance programs; (4) coordinating and evaluating cooperating states' food safety activities; and (5) developing and disseminating food safety and regulatory information to consumers and industry. FDA's Center for Veterinary Medicine (CVM) is responsible for ensuring that all animal drugs, feeds (including pet foods), and veterinary devices are safe for animals, are properly labeled, and produce no human health hazards when used in food-producing animals.

The FDA also cooperates with over 400 state agencies across the nation that carry out a wide range of food safety regulatory activities. However, the state agencies are primarily responsible for actual inspection. FDA works with the states to set the safety standards for food establishments and commodities and evaluates the states' performance in upholding such standards as well as any federal standards that may apply. FDA also contracts with states to use their food safety agency personnel to carry out certain field inspections in support of FDA's own statutory responsibilities.

Food Safety and Inspection Service

FSIS regulates the safety,wholesomeness, and proper labeling of most domestic and imported meat and poultry and their products sold for human consumption. Under the Federal Meat Inspection Act of 1906, as amended (21 U.S.C. 601 et seq.), FSIS inspects all cattle, sheep, swine, goats, and equines during slaughtering and processing. Under the Poultry Products Inspection Act of 1957, as amended (21 U.S.C. 451 et seq.), FSIS is required to inspect "any domesticated bird" being processed for human consumption; however, USDA regulations implementing this law limit the definition of domesticated birds to chickens, turkeys, ducks, geese, ratites (emus, ostriches, and rheas), and guineas. FDA has jurisdiction over exotic and alternative meats not inspected by FSIS, and shares the responsibility for egg safety with FSIS. The latter is responsible for the safety of liquid, frozen, and dried egg products, domestic and imported, and for the safe use or disposition of damaged and dirty eggs under the Egg Products Inspection Act, as amended (21 U.S.C. 1031 et seq.).

FSIS staff numbers around 9,400; roughly 8,000 of them, including about 1,000 veterinarians, are in about 6,300 meat slaughtering and/or processing plants nationwide. FSIS personnel inspect all meat and poultry animals at slaughter on a continuous basis, and at least one federal inspector is on the line during all hours the plant is operating

Processing inspection does not require an FSIS inspector to remain constantly on the production line or to inspect every item. Instead, inspectors are on site daily to monitor the plant's adherence to the standards for sanitary conditions, ingredient levels, and packaging, and to conduct statistical sampling and testing of products. Because all plants are visited daily, processing inspection also is considered to be continuous.

FSIS also is responsible for certifying that foreign meat and poultry plants are operating under an inspection system equivalent to the U.S. system before they can export their product to the United States. FSIS inspectors located at U.S. ports of entry carry out a statistical sampling program to verify the safety of imported meats from cattle, sheep, swine, goats, and equines and imported poultry meat from chickens, turkeys, ducks, geese, quail, ratites, and guineas before they are released into domestic commerce. FDA is responsible for ensuring the safety of imported meat from any other species.

Approximately 28 states operate their own meat and/or poultry inspection programs. FSIS is statutorily responsible for ensuring that the states' programs are at least equal to the federal program. Plants processing meat and poultry under state inspection can market their products only within the state. If a state chooses to discontinue its own inspection program, or if FSIS determines that it does not meet the agency's equivalency standards, FSIS must assume the responsibility for inspection if the formerly state-inspected plants are to remain in operation. FSIS also has cooperative agreements with more than two dozen states under which state inspection personnel are authorized to carry out federal inspection in meat and/or poultry plants. Products from these plants may travel in interstate commerce.[6]

Centers for Disease Control and Prevention (HHS)

CDC is responsible for (1) monitoring, identifying, and investigating foodborne disease problems to determine the contributing factors; (2) working with FDA, FSIS, NMFS, state and local public health departments, universities, and industry to develop control methods; and (3) evaluating the effect of control methods. In 1995, CDC launched 'FoodNet,' a collaborative project with the FDA and USDA to improve data collection on foodborne illness outbreaks. FoodNet includes active surveillance of clinical microbiology laboratories to obtain a more accurate accounting of positive test results for foodborne illness; a physician survey to determine testing and laboratory practices; population surveys to identify illnesses not reported to doctors; and research studies to obtain new and more precise information about which food items or other exposures may cause diseases. FoodNet data allows CDC to have a clearer picture of the incidence and causes of foodborne illness and to establish baseline data against which to measure the success of changes in food safety programs. The Public Health Service Act, as amended (42 U.S.C. 201 et seq.), provides the legislative authority for CDC's food safety related activities.

National Marine Fisheries Service (DOC)

Although the FDA is the primary agency responsible for ensuring the safety, wholesomeness and proper labeling of domestic and imported seafood products, NMFS conducts, on a fee-for-service basis, a voluntary seafood inspection and grading program that focuses on marketing and quality attributes of U.S. fish and shellfish. The primary legislative authority for NMFS's inspection program is the Agricultural Marketing Act of 1946, as amended (7 U.S.C. 1621 et seq.). NMFS has approximately 160 seafood safety and quality inspectors, and inspection services are funded with user fees.

Environmental Protection Agency

EPA has the statutory responsibility for ensuring that the chemicals used on food crops do not endanger public health. EPA's Office of Pesticide Programs is the part of the agency that (1) registers new pesticides and determines residue levels for regulatory purposes; (2) performs special reviews of pesticides of concern; (3) reviews and evaluates all the health data on pesticides; (4) reviews data on pesticides' effects on the environment and on other species; (5) analyzes the costs and benefits of pesticide use; and (6) interacts with EPA regional offices, state regulatory counterparts, other federal agencies involved in food safety, the public, and others to keep them informed of EPA regulatory actions. The Federal Insecticide, Fungicide, and Rodenticide Act, as amended (7 U.S.C. 136 et seq.), and the Federal Food, Drug, and Cosmetic Act, as amended (21 U.S.C. 301 et seq.), are the primary authorities for EPA's activities in this area.

Other Federal Agencies with Food Safety Responsibilities

Among the other agencies that play a role in food safety, USDA's Agricultural Research Service (ARS) performs food safety research in support of FSIS's inspection program. It has scientists working in animal disease bio-containment laboratories in Plum Island, NY, and Ames, IA. USDA's Animal and Plant Health Inspection Service (APHIS) indirectly protects the nation's food supply through programs to protect plant and animal resources from domestic and foreign pests and diseases, such as brucellosis and bovine spongiform encephalopathy (BSE, or "mad cow" disease). The Department of Homeland

Security (DHS) is responsible for coordinating agencies' food security activities, including at U.S. borders.

Congressional Committees

In the Senate, food safety issues are considered by the Committees on Agriculture, Nutrition, and Forestry; Homeland Security and Governmental Affairs; and Health, Education, Labor, and Pensions. In the House, various food safety activities fall under the jurisdiction of the Committees on Agriculture; Energy and Commerce; Oversight and Government Reform; and Science. Agriculture subcommittees of the House and Senate Appropriations Committees also serve oversight and funding roles in how the major agencies carry out food safety policies.

Reorganization of the Federal Food Safety Regulatory Structure

In its annual (January 2007) report, GAO designated food safety oversight as one of 29 "high risk" federal program areas. The report stated: "Any food contamination could undermine consumer confidence in the government's ability to ensure the safety of the U.S. food supply, as well as cause severe economic consequences. The current fragmented federal system has caused inconsistent oversight, ineffective coordination, and inefficient use of resources. GAO has recommended that Congress consider a fundamental reexamination of the system and other improvements to help ensure the rapid detection of and response to any accidental or deliberate contamination of food before public health and safety is compromised."[7]

Earlier, the Institute of Medicine of the National Academy of Sciences (IOM/NAS) and the National Commission on the Public Service studied the issue and recommended options for change.[8] IOM/NAS has been among those who maintain that reorganizing agencies, by itself, would not provide safer food; its 1998 report recommended a more science-based statute and additional inspection resources, among other things. Opponents of major food safety changes, including many in the food and agricultural industries, assert that the system already is scientifically based, that the statutes are adequate, and that food companies already produce and distribute safe food, making the U.S. system a model for food safety around the world.

Legislative proposals to combine these agencies and responsibilities into a single federal food safety agency were introduced in past Congresses, but did not advance. In the 1 10th Congress, the first bills to establish a new Food Safety Administration were offered on February 15, 2007, as H.R. 1148 by Representative DeLauro and S. 654 by Senator Durbin. On February 8, 2007, the House Appropriations Subcommittee on Agriculture, Rural Development, Food and Drug Administration, and Related Agencies, which is chaired by Representative DeLauro, held a hearing on food safety reorganization.

Another bill that would alter some agency food roles is S. 1082, the Senate-passed FDA Revitalization Act, which includes an amendment requiring establishment of a new FDA-administered food safety notification system. A similar provision is part of two broader companion measures, the Human and Pet Food Safety Act of 2007 (S. 1274/H.R. 2108). Also, H.R. 1600, H.R. 2629, and S. 887 propose to return agricultural border inspectors to APHIS, where they resided before their responsibilities were moved to the Department of Homeland Security.[9]

Lawmakers may be asked to consider these or other proposals that would either reorganize or consolidate the federal food safety organization. A range of policy options could be debated, including whether the current regulatory approaches and their authorizing statutes remain appropriate, particularly given the diversity of food types, different health risks, methods of production, and sources of supply; the continuously evolving science on foodborne illness and how to prevent future outbreaks; the impacts on industry competitiveness, particularly in a global economy; and funding constraints.

REFERENCES

[1] Background on the agencies is based largely on CRS Report 98-91, *Food Safety Agencies and Authorities: A Primer*, by Jean M. Rawson, Alejandro E. Segarra, and Donna U. Vogt (out of print). Primary sources for that report included various documents and materials provided by federal food safety agencies and by the U.S. Government Accountability Office.

[2] [2] Roughly two-thirds of the $1 trillion was for domestically produced farm foods; imports and seafood account for the balance. Source: USDA, Economic Research Service, data accessed February 2007 at the "Food Sector" Web page at [http://www.ers.usda.gov /Browse/FoodSector/].

[3] U.S. Department of Health and Human Services, Centers for Disease Control and Prevention, *Foodborne Illness: Frequently Asked Questions*, at [http://www.cdc.gov/foodsafety/].

[4] *High Risk Series: An Update* (GAO-07-310), January 2007.

[5] *Agriculture, Rural Development, Food and Drug Administration, and Related Agencies Appropriations for 2007*, hearings before a Subcommittee of the Committee on Appropriations, House of Representatives, 109th Cong., 2nd sess. (Part 5, pp. 755 and 848).

[6] See also CRS Report RL32922, *Meat and Poultry Inspection: Background and Selected Issues*, by Geoffrey S. Becker.

[7] GAO, *High Risk Series: An Update*. Numerous earlier GAO reports reiterated this theme.

[8] See National Research Council, Institute of Medicine, *Ensuring Safe Food From Production to Consumption*, Washington, DC, National Academy Press, 1998; and National Commission on the Public Service, *Urgent Business For America: Revitalizing the Federal Government For the 21st Century*, Washington, DC, 2003.

[9] See also CRS Report RL32521, *Agroterrorism Threats and Preparedness*, by Jim Monke; and CRS Report RS22664, *U.S. Food and Agricultural Imports: Safeguards and Selected Issues*, by Geoffrey S. Becker.

In: Food Protection and Safety
Editor: Lyman F. Thomasson

ISBN 978-1-60692-187-6

Chapter 5

PRODUCE SAFETY FROM PRODUCTION TO CONSUMPTION: 2004 ACTION PLAN TO MINIMIZE FOODBORNE ILLNESS ASSOCIATED WITH FRESH PRODUCE CONSUMPTION*

U.S. Food and Drug Administration

BACKGROUND

The Challenge of Foodborne Illness:

Foodborne illness continues to present a public health challenge in the United States (U.S.). The Centers for Disease Control and Prevention (CDC) estimates that, in the U.S. each year, 76 million people become sick, more than 325,000 people are hospitalized, and 5,000 people die from foodborne illness. The estimated economic cost of foodborne illness in terms of pain and suffering, reduced productivity, and medical expenses is substantial, in the range of $ 10-83 billion each year. Foodborne illness is associated with a range of foods, including fresh produce.

Produce is recognized as an important component of a healthy diet because it is a source of vitamins, minerals, fiber, and antioxidants. Produce can play an important role in weight management as well. Because most produce is grown in a natural environment, it is vulnerable to contamination with pathogens. Factors that may affect the occurrence of such contamination include agricultural water quality, the use of manure as fertilizer, the presence of animals in fields or packing areas, and the health and hygiene of workers handling the produce during production, packing, processing, transportation, distribution, or preparation. The fact that produce is often consumed raw without any type of intervention that would reduce, control, or eliminate pathogens prior to consumption contributes to its potential as a source of foodborne illness.

CDC estimates that, in the 1990's, at least 12 percent of foodborne-outbreak-associated illnesses were linked to fresh produce items. Over the past decade, the federal government has focused significant resources on reducing foodborne illness from all sources. However, despite these efforts, foodborne illness associated with fresh produce continues to be documented. The persistence of foodborne illness associated with fresh produce may be attributable to a number of factors, but many cases are preventable. Given the importance of produce consumption and its central role in a healthy diet, it is imperative that the incidence of foodborne illness cases associated with produce be reduced.

As the federal agency principally responsible for regulating the safety of the majority of the food supply, including produce, the Food and Drug Administration (FDA) is prepared to lead the effort to achieve this reduction. Realistically, however, reducing foodborne illness associated with fresh produce consumption will require a collaborative effort by FDA, its federal food safety partners such as the CDC and the U.S. Department of Agriculture (USDA), FDA's counterparts in foreign governments, state and local agencies, the private sector (including relevant trade associations), academia and consumers.

Goal and Objectives of the Produce Safety Action Plan (Action Plan):

The overarching goal of FDA's Action Plan is to minimize the incidence of foodborne illness associated with the consumption of fresh produce. To achieve this goal, the Action Plan has four general objectives: 1) Prevent Contamination of Fresh Produce with Pathogens; 2) Minimize the Public Health Impact When Contamination of Fresh Produce Occurs; 3) Improve Communication with Producers, Preparers, and Consumers about Fresh Produce; and 4) Facilitate and

* Excerpted from U.S. FDA report dated October 2004.

Support Research Relevant to Fresh Produce. For each objective, FDA's Action Plan identifies steps that could contribute to the achievement of the objective.

FDA believes that the most effective strategy for reducing foodborne illness from fresh produce is one that approaches the problem from several different angles. For this reason, the Action Plan anticipates that FDA's food safety partners in both the public and private sectors will participate in the identified activities. By working both independently and cooperatively, these agencies or groups will ensure maximum progress toward the goal of reducing the incidence of foodborne illness associated with the consumption of fresh produce.

Scope of the Action Plan:

FDA's Action Plan is designed to target microbial food safety hazards (such as bacteria, viruses, and parasites) in or on produce consumed in the U.S., whether produced in the U.S. or abroad. FDA believes that each entity involved in producing, packing, processing, transporting, distributing, or preparing fresh produce has a responsibility to conduct its activities so as to reduce, control, or eliminate microbial contamination of produce. Thus, the Action Plan extends to all parts of the food chain from farm through retail or consumer preparation and consumption and is intended to cover fresh fruits and vegetables, both in their unpeeled, natural form and raw products that have received some minimal processing (such as peeling, chopping, or trimming). The Action Plan is not intended to cover frozen fruits and vegetables, fruit and vegetable juices, or other commodities such as tree nuts that are neither fruits nor vegetables and not typically regarded as produce.

Measuring the Success of the Action Plan:

FDA recognizes that developing this Action Plan is only one of many steps required to achieve the overall goal of reducing foodborne illness associated with the consumption of fresh produce. Implementation is, of course, critical to the success of any action plan. Likewise, it is essential to evaluate the effects of implementation. Thus, FDA intends to measure the effects of the Action Plan. An overall measure of the success of the Action

Plan would be a decrease in the number of foodborne illness outbreaks, and the number of illnesses per outbreak associated with fresh produce. FDA will work closely with CDC to monitor the incidence of foodborne illnesses associated with fresh produce consumption. The agency acknowledges, however, that this is not a completely reliable indicator because of many factors. The success of each individual objective in the plan can be measured by the following:

- Objective 1 can be measured by the issuance of guidance, regulations and tools to assist implementation and an assessment of the adoption of guidance or adherence to regulations by relevant producers, and by the incidence of contaminated produce identified during sampling surveys.
- Objective 2 can be measured by the speed and accuracy of epidemiological, traceback, and environmental investigations, and the information gained from such investigations.
- Objectives 3 and 4 can be measured by whether the action items have been successfully accomplished.

OBJECTIVES OF THE ACTION PLAN

Objective 1. Prevent Contamination of Fresh Produce with Pathogens

This objective can be achieved by steps such as the following:

Guidance and Regulations

- Periodically review FDA's "Guide to Minimize Microbial Food Safety Hazards for Fresh Fruits and Vegetables" and revise to reflect advances in science or technology and findings from foodborne illness outbreaks and other investigations
- Develop, and assist in the development of, additional guidance, e.g., commodity-specific and practice- specific guidance relevant to producing, packing, processing, transporting, distributing, or preparing fresh produce (such as guidance for safe processing of fresh-cut produce)
- Develop guidance to promote improved hygiene and better preparation practices in the retail environment
- Develop commodity-specific handling advice for preparers of fresh produce, such as retailers, restaurants, food service operations, and

consumers, and share the advice with state and local food safety agencies for further dissemination

- Initiate rulemaking to minimize foodborne illness associated with the consumption of sprouted seeds
- Add language to the FDA Food Code, as needed, to address produce safety at retail
- Assess the extent of the adoption of guidance and adherence to regulations by relevant sectors in the fresh produce supply chain

Education Outreach

- Promote the application of Good Agricultural Practices and Good Manufacturing Practices (GAPs and GMPs) to fresh produce production, using FDA's "Guide to Minimize Microbial Food Safety Hazards for Fresh Fruits and Vegetables" or comparable guidance emphasizing producers that have not been previously reached
- Promote the application of other guidance for fresh produce production (such as FDA's two guidance documents about growing sprouts, "Reducing Microbial Food Safety Hazards For Sprouted Seeds and Sampling" and "Microbial Testing Of Spent Irrigation Water During Sprout Production")
- Raise and maintain consumer awareness about how to handle fresh produce safely by periodic distribution of information through community outreach programs, in the media (print and non-print), and by other means
- Share educational outreach and consumer advisory materials with federal, state, and local agencies, foreign governments, and the private sector for broad dissemination
- Use the internet, such as the federal government's food safety website (www.foodsafety.gov), to promote ready access to educational materials for fresh produce producers, distributors, retailers, food workers, and consumers

Facilitating Implementation

- Develop, support, and promote tools (e.g., fact sheets, model forms, Standard Operating Procedures (SOPs), checklists) to assist in the implementation of guidance and regulations that cover segments in the fresh produce supply chain
- Identify and promote requirements of other legal authorities that support practices and conditions to prevent the microbial contamination of fresh produce (e.g., Occupational Safety and Health Act and 29 CFR 1928.110, field sanitation laws and regulations that establish the minimum ratio of toilets to workers and proper hand washing facilities, maximum worker to restroom distance and maintenance of facility cleanliness)
- Identify and seek solutions to requirements of other legal authorities that could inadvertently increase the risk of microbial contamination of fresh produce (e.g., state water conservation requirements)

Objective 2. Minimize the Public Health Impact when Contamination of Fresh Produce Occurs

This objective can be achieved by steps such as the following:

- Use tools such as facility inspections and surveys, increase routine monitoring of practices and conditions in the fresh produce supply chain, including farms, packing facilities, and distribution centers, to identify practices that could lead to the spread of contamination in produce
- Expand surveillance of fresh produce, including both focused sampling of fresh produce with a history of association with illness outbreaks and expanded sampling of a broader range of produce items to obtain baseline data on the incidence of contamination
- Continue to improve and refine appropriate regulatory follow up when insanitary conditions are identified or contamination of fresh produce is found
- Increase the speed of epidemiological, traceback, and environmental investigations and the quality of the information obtained through such investigations
- Review and revise, as appropriate, the criteria for initiating traceback and farm investigations and the composition of investigative teams

- Prepare training or guidance as needed for epidemiological, traceback, and environmental investigations by federal, state, and local investigators to promote consistency of investigations and identification of potential sources of contamination and to promote preventive measures to ensure produce safety
- Support the collection and analysis of outbreak surveillance data by CDC to help assess the impact of the Action Plan
- Develop systems for exchange and prompt use of data between all federal, state, and local food safety agencies that conduct epidemiological investigations of foodborne illness outbreaks and to ensure the prompt utilization of the results of such investigations
- Enhance PulseNet capacity to expedite identification and comparison of foodborne pathogens likely to be transmitted by produce
- Collaborate with international organizations, such as the World Health Organization and the Pan American Health Organization, to develop and sustain PulseNet capacity in other countries that export fresh produce to the U.S.

Objective 3. Improve Communication with Producers, Packers, Processors, Transporters, Distributors, Preparers, Consumers, and Other Government Entities about Fresh Produce

This objective can be achieved by steps such as the following:

- Promote improved lines of communication between federal, state, and local food safety agencies, foreign governments, the public and the private sector on an ongoing basis and in response to a specific event, such as a foodborne illness outbreak or other crisis associated with fresh produce
- Promote transparency in the development of guidance, regulations, and other protocols designed to address fresh produce safety
- Encourage the two-way exchange of information between FDA and all stakeholders throughout the supply chain on surveys, research, investigational findings, and industry practices throughout the supply chain

- Collaborate with appropriate stakeholders to gain information on industry practices associated with implicated product in order to focus traceback, facilitate investigations, and maintain consumer confidence in the food supply
- Establish a protocol, with stakeholder input, to ensure that consumers are informed as quickly as possible about a foodborne illness outbreak involving an FDA-regulated product and share the protocol with stakeholders
- Establish mechanisms to ensure prompt communication between federal, state, and local food safety agencies, foreign governments, and the private sector when there is a concern about a particular region, practice, or produce commodity group.

Objective 4. Facilitate and Support Research Relevant to the Contamination of Fresh Produce

This objective can be achieved by steps such as the following:

- Identify research likely to make the most significant contributions to reducing the public health risk from contaminated fresh produce, with a focus on approaches that provide practical information
- Develop risk-based approaches directed at preventing contamination of fresh produce or identify effective interventions to address contamination that has occurred
- Assess relative risk for specific areas of concern for fresh produce (such as agricultural and processing water quality, environmental contamination during production, unsafe handling practices, insanitary equipment, employee health and hygiene, and the practices that hinder traceback) and clarify mechanisms of contamination under commercial conditions in produce supply chain settings
- Develop additional sampling protocols, as needed, to facilitate efficient and reliable detection of pathogens in or on fresh produce and in the fresh produce production environment
- Develop revised or new analytical methods for improving the timely detection of pathogens in or on fresh produce and in the fresh produce production environment, with a focus on the commodity/pathogen combinations of greatest public health concern

- Leverage research resources aimed at developing sampling and pathogen detection methods for fresh produce through cooperative efforts between federal agencies, state and local offices, academic institutions, and the private sector
- Assess the efficacy of guidance, educational outreach, and training materials and validate the efficacy of industry practices and interventions
- Identify barriers to implementing best practices in the production, processing, and preparation of fresh produce to assist in the development of effective intervention strategies
- Promote the technology transfer of research findings.

In: Food Protection and Safety
Editor: Lyman F. Thomasson
ISBN 978-1-60692-187-6

Chapter 6

FDA FOOD PROTECTION PLAN SIX-MONTH PROGRESS SUMMARY*

U.S. Food and Drug Administration

In November 2007, the Food and Drug Administration released the Food Protection Plan (FPP), an integrated strategy for protecting the nation's food supply. The FPP addresses both food safety and food defense for domestic and imported products. FDA is working collaboratively across the agency to address the three core elements of protection: prevention, intervention and response. This summary provides an update on the accomplishments under the FPP in the first six-months and an overview of activities planned to begin now or in the near future.

OVERVIEW

Prevention

Several key prevention initiatives have been started since November 2007. These began with outreach to explain and involve others in FDA's Food Protection Plan. This outreach has involved multiple meetings with various

* Excerpted from U.S. FDA website http://www.fda.gov/oc/initiatves/advance/food/progressreport.html.

foreign countries, state and local organizations, and industry and consumer groups. To further the interaction with federal, state and local partners, FDA announced a meeting to share information and develop strategies and future activities between federal, state, and local partners. As a cross-cutting activity, FDA has formed a Risk- Based Steering Committee with the charge of focusing FDA's risk-based prevention, intervention and response activities to focus on the needs, processes, and timelines to ensure that a comprehensive risk-based approach is taken with regard to food protection. Specific risk-based prevention activities include FDA working in collaboration with states, universities and industry on a Tomato Safety Initiative. In an effort to increase foreign capacity and FDA's presence beyond our borders, FDA has engaged with India and begun implementation of the China Memorandum of Agreement. The first bilateral meeting with China was held in Beijing in March, 2008.

Intervention

FDA is working to improve the targeting of risk-based inspections and sampling to focus resources where they will have the greatest impact. FDA has completed a three-year plan to increase state inspections and will hire an additional 130 employees to conduct food field exams, inspections, and sample collections. FDA issued a Federal Register notice to solicit input on third-party certification programs and is interacting with a variety of industry groups to explore this further. FDA is working to develop better tools to identify food safety threats at the border, including the development of an advanced screening system. This system, known as PREDICT, was piloted on a small scale in recent months and the assessment of the pilot is currently under agency review. FDA has established a research coordinating committee to develop a collaborative research agenda that supports activities under prevention, intervention and response, such as mitigation strategies and rapid detection systems.

Response

In order to improve the response to a threat to food safety, FDA is engaging with stakeholders to identify traceability best practices to be able to quickly identify where a contaminated product came from and where it has been distributed. FDA is working with states to improve response through Incident Command System training and is preparing to issue a Request for Applications for states interested in establishing Rapid Response Teams to enable rapid, localized response to incidents. In an effort to improve communication with the public, FDA developed templates for recall communication and presented them to FDA's Risk Communication Advisory Board for input in March 2008.

Prevention

1.1 Promote Increased Corporate Responsibility to Prevent Foodborne Illness

Outreach to States, Industry, and Other Stakeholders

Accomplishments

FDA has undertaken several outreach activities in the last six months.

- FDA has begun working with state and local partners on developing a strategy for the implementation of the Food Protection Plan (FPP). The Commissioner announced that a 50-state meeting will be held on August 12-14, 2008 in St. Louis, Missouri to share information and develop strategies and future activities between federal, state, and local partners and FDA for the FPP, Import Safety Action Plan (ISAP), and Food and Drug Administration Amendments Act of 2007 (FDAAA).
- FDA issued a Federal Register Notice announcing a docket for stakeholder comment on the implementation of the FPP as part of a broad outreach plan. Docket number FDA-2008-N-0188, Food Protection Plan; Outreach Activities; Opportunity for Public Comment, can be found at www.regulations.gov.
- FDA issued a Federal Register Notice for a public meeting to solicit input on pet food labeling and standards for ingredients and processing of animal feed. Docket number FDA-2007-N-0442, Meeting Being Planned to Obtain Public Input for Ensuring the Safety of Pet Food, can be found at www.regulations. gov.
- FDA issued a Federal Register Notice for a public meeting to present changes to the Animal Feed Safety System project and the ranking of feed hazards according to the risk they pose to animal and public health. Docket number FDA-2008-N-0189-0001, Meeting to Present (1) Changes to the Animal Feed Safety System (AFSS) Project and (2) The

Ranking of Feed Hazards According to the Risks They Pose to Animal and Public Health; Part 3: Swine Feed Example; Public Meeting, can be found at www. regulations.gov.

Looking Ahead

- FDA will be holding bilateral meetings in the next few months with Health Canada and the Canadian Food Inspection Agency, Mexico's Federal Commission for the Protection Against Sanitary Risk (COFEPRIS) and Mexico's Secretary for Agriculture, Ranching, Rural Development, Fisheries, and Food Supply (SAGARPA).

Building in Safety Upfront by Identifying Best Practices and Standards

Accomplishments

FDA is working with industry and stakeholders to identify and ensure the use of best practices for early prevention of food safety threats.

- FDA is working in collaboration with the State Health and Agriculture departments in Virginia and Florida, several universities, and the produce industry on a multi-year Tomato Safety Initiative. As part of the initiative, FDA has led assessments of grower practices focusing on the factors believed to be associated with contamination of tomatoes with Salmonella. FDA has conducted assessments in Virginia and completed two sets of assessments in Southern and Central Florida. FDA is in the process of analyzing the information in order to prepare a report on the assessments that can be shared with industry and other stakeholders.
- FDA has gathered input on animal feed best practices for preventive controls from industry groups including the National Grain and Feed Association, the American Feed Industry Association, the Pet Food Institute, and the Association of American Feed Control Officials (an organization of state feed control officials).
- FDA issued a draft Compliance Policy Guide (CPG) on Listeria monocytogenes. The CPG provides guidance for FDA staff on the agency's enforcement policy for Listeria monocytogenes in ready-to-eat food. FDA issued a Federal Register Notice to solicit comment and held public hearings on the draft. Docket number FDA-2008-D-0058-0001,

Draft Compliance Policy Guide Sec.555.320 Listeria monocytogenes, can be found at www.regulations.gov. FDA also issued draft guidance on controls that processors can use to minimize contamination of food with Listeria monocytogenes.

- FDA released self-assessment tools for industry to minimize the risk of intentional contamination of food and cosmetics.
- FDA issued a final rule entitled, "Substances Prohibited from Use in Animal Food or Feed." This rule enhances the current Bovine Spongiform Encephalopathy (BSE) rule by prohibiting certain specified risk materials in all animal feed in addition to all of the original requirements. Docket number 2002-N-0031 can be found at www.regulations.gov.

Looking Ahead

FDA is working on developing, modifying and updating several food safety documents to address the current food safety vulnerabilities.

- FDA will issue a Federal Register notice to solicit input on possible modification of industry food safety guidelines for leafy greens and melons as part of FDA's consideration of adopting the guidance.
- FDA will issue a Federal Register notice to solicit input on updating the 1998 Good Agricultural Practices (GAPs).
- FDA intends to publish a Final Rule on requirements to prevent Salmonella enteritidis contamination of shell eggs during egg production.
- FDA plans to release the 4th Edition of Fish and Fishery Products Hazards and Controls Guidance with updates to the previous editions to incorporate the current scientific and technical information regarding hazards associated with the harvest, processing, and storage of fish and fishery products.
- FDA is working on modernizing the Current Good Manufacturing Practices (GMPs) for human food. There have been changes in both the food industry and in the science of food safety that indicate a need for modernization.
- FDA will be querying key counterparts to identify best practices for import safety, including the use of third-party inspectors and reliance on certificates.

Increasing Foreign Capacity and Technical Assistance Accomplishments

FDA is building foreign capacity by engaging with foreign partners on both the FPP and ISAP. In addition, FDA is providing technical assistance to foreign countries to ensure the safety of imported food.

- A group from FDA attended the Food Safety Quadrilateral meeting with Canada, New Zealand and Australia in April 2008 to advance the FPP and ISAP. One of the proposals under consideration is the establishment of a rapid alert system to share information from any of the countries regarding potential significant public health problems found in food or feed.
- In response to finding unapproved drugs in Chinese aquaculture products, FDA conducted a review of aquaculture drug residue control procedures at thirteen Chinese firms that have been certified by the Chinese government as in compliance with the requirements of the import alert on Chinese aquaculture products. FDA plans to conduct verification audits of these firms in early July 2008.
- FDA was part of a Department of Health and Human Services delegation to Vietnam to discuss establishing a cooperative arrangement with the Ministry of Health covering food, feed, and medical products. An agreement with Vietnam was recently signed.
- FDA provided a briefing to Washington, DC based foreign embassy personnel on FDA actions under the FPP and ISAP.

Looking Ahead

- FDA is scheduling meetings with counterparts in Canada and Mexico to reinvigorate our bilateral relationships and to step up our collaborations to advance the work under the FPP and ISAP.

Expanding FDA Beyond our Borders

Accomplishments

FDA has engaged with China and India on expansion of FDA presence beyond our borders.

- Implementation of the China Memorandum of Agreement (MOA) has begun. FDA has met their first set of deadlines by providing registration materials to the Chinese government, identifying points of contact for the MOA, and drafting the first five-year work plan.
- The first bilateral meeting with China was held in March 2008 in Beijing. The meeting focused on the relationship with the General Administration of Quality, Supervision, Inspection, and Quarantine (AQSIQ). There was verbal agreement to limit the present efforts in fulfilling the MOA to aquaculture (five species plus tilapia) and ingredients (wheat gluten, corn gluten, and rice protein).
- An FDA delegation visited Indian counterparts to discuss requirements for an FDA presence in India. Results are promising for future collaboration.
- In April, the Commissioner traveled to Jordan to explore the possibilities of enhanced collaboration between FDA and Jordan.

Looking Ahead

- FDA is exploring an FDA presence in Latin America/Central America.
- FDA is developing the animal feed requirements criteria for animal feed products covered under the MOA with China.

1.2. Identify Food Vulnerabilities and Assess Risks

Identifying Food Vulnerabilities

Accomplishments

Research has been conducted in a number of areas related to both food safety and food defense. A better understanding of the stability of contaminants in food leads to a better understanding of potential threats to food safety.

- FDA assessed and published data on the microbiological load of bagged, ready-to-eat produce. These findings suggest that in bagged, readyto-eat lettuce, the microbial load may be relatively higher than expected. FDA published these findings in February 2008 in the Journal of Applied Environmental Microbiology. FDA is planning a follow-up study.

- FDA completed an Inter-Agency Agreement with the United States Department of Agriculture (USDA) and the Department of Homeland Security (DHS) to determine the survivability of Bacillus anthracis (anthrax) in processed liquid egg products, which include whole eggs, egg yolks, and egg whites. Further studies are being conducted to determine the role of lyzozyme in Bacillus anthracis inactivation.
- FDA developed an assay to assess the stability of two bioterrorism agents in high-risk foods. This assay can be used to assess other chemicals that may be used by terrorists to contaminate the food supply.

Looking Ahead

- FDA is researching new ways to conduct molecular epidemiological analysis of Salmonella newport. In addition, scientists have identified several natural plant bacteria that are effective in preventing Salmonella newport from contaminating commercial tomato fruit.

Developing Risk-Based Processes Accomplishments

FDA established a Risk-Based Steering Committee to focus on the needs, processes, and timelines to ensure that a comprehensive risk-based approach is taken with regard to food protection.

- FDA's Office of Regulatory Affairs (ORA), Center for Food Science and Nutrition (CFSAN), and the Food Field Committee collaborated to develop and use a risk-based high priority list to create a FY 2008 work plan for the ORA field force.
- FDA's Center for Veterinary Medicine established a team to enhance modeling capability for relative risk-ranking across programs.

Looking Ahead

As part of the risk-based approach, FDA will use the information gathered during the FY 2008 risk-based inspections to assess the risk- ranking models and to determine which models need fine tuning and which additional models may need to be developed, and will work towards developing a system for relative risk ranking.

1.3. Expand the Understanding and Use of Effective Mitigation Measures

Expanding FDA Research, Development, and Evaluation

Accomplishments

Several research projects on mitigation efforts are currently underway.

- FDA approved the use of cetylpyridinium chloride (CPC) as an antimicrobial agent applied in solution to raw poultry carcasses to prevent contamination with Salmonella or E. coli.
- FDA is using genetic analysis to identify hundreds of Salmonella strains from seafood imports. The analysis provides information that can be used to trace Salmonella outbreaks and implement surveillance programs to ensure food safety.

Looking Ahead

- FDA has established a research coordinating committee and is developing a collaborative and integrated FPP research agenda that will address research needs to support protection, intervention and response.
- FDA has initiated a collaborative multi-institutional study to reduce the risk of E. coli O1 57:H7, funded by USDA under the National Integrated Food Safety Initiative. The work will examine pathogen risk mitigation strategies for leafy greens from field to table.
- FDA plans to conduct research on the susceptibility of pathogens found in raw and processed meats and imported seafood to antimicrobial agents and mechanisms by which these pathogens develop a resistance to antimicrobial agents.
- FDA is partnering with outside partners on research, development and evaluation of mitigation tools. FDA established a Memorandum of Understanding with DHS and the Department of Justice (DOJ) to develop forensic tools to allow the identification and differentiation of individual strains of foodborne bacteria. This will assist in rapid identification of the source of contamination.

Intervention

2.1. Focus Inspection and Sampling based on Risk

Expanding Field Capacity through Risk-Based Inspections and Sampling

Accomplishments

Risk-based inspections allow FDA to focus resources where they will have the greatest impact.

- A targeted risk-based inspection of a canning facility identified cans with viable Clostridium botulinum spores and a recall was initiated. FDA initiated this inspection, along with inspections of other low acid canned food (LACF) manufacturers, following four cases of botulism in consumers in 2007. FDA increased inspection efforts to ensure that manufacturers of all types of LACF products are adhering to applicable FDA requirements. These actions illustrate the need for companies to operate under adequate preventive control systems.
- ORA has completed a 3-year plan for increasing state inspections.

Looking Ahead

- Based on the FY 2009 President's Budget, FDA plans to conduct an additional 327 state contract food inspections in FY 2009 over the FY 2008 estimate.
- FDA is considering additional targeted assignments to other segments of the LACF industry in light of findings of viable Clostridium botulinum spores at a canning facility.
- FDA will hire and train at least 161 new FTEs to conduct food safety work. The Office of Regulatory Affairs will hire at least 130 new FTEs to conduct food field exams, inspections, and sample collections. The Center for Food Safety and Applied Nutrition will hire at least 29 FTEs and the Office of Crisis Management will hire at least two new FTEs using FY 2008 appropriated dollars.
- Based on the FY 2009 President's Budget, in FY 2009, ORA will conduct an additional 20,000 food import field exams above the FY 2008 performance goal.

Ensuring Import Safety though Voluntary Third-Party Certification Programs

Accomplishments

- FDA issued a Federal Register Notice for Food and Feed Third-Party Certification Programs. The notice seeks public comment on the existence and use of third-party certification programs to better understand how they can be used help to ensure that food products are safe, secure, and meet FDA requirements. Docket Number FDA-2008-N-0183-001, Third-Party Certification Programs for Foods and Feeds, can be found at www.regulations.gov.

Looking Ahead

FDA is exploring current existing third-party certification programs to gain a better understanding of the general attributes of such programs to ensure certification is reliable.

2.2. Enhance Risk-Based Surveillance of Imported Foods at the Border

Advancing Screening Technology at the Border

Looking Ahead

- FDA is working to develop better tools to identify food safety threats at the border. Under contract to FDA, New Mexico State University is developing a prototype system for improving electronic screening, using open-source intelligence, of imported products offered for entry into the U.S. The evaluation of the prototype system, PREDICT (Predictive Risk-Based Evaluation of Dynamic Import Compliance Targeting), has been completed and the final pilot evaluation document is currently under review.
- FDA plans to publish the Prior Notice of Imported Foods Final Rule by the end of 2008.

2.3. Improve the Detection of Food Systems "Signals" that Indicate Contamination

Deploy New Rapid Detection Tools to Identify Pathogens

Accomplishments

FDA is developing new methods for rapidly and accurately identifying contaminants in food.

- FDA has developed a rapid detection method using flow cytometry to identify E. coli and Salmonella in food. This system is being used in poultry processing facilities to detect and prevent bacterial contamination during food processing.
- FDA microbiologists received training at CDC's Salmonella Reference Laboratory on a new molecular method for rapidly and accurately identifying Salmonella serovars. The instruments have been purchased by both CFSAN and ORA laboratories.
- FDA has developed a method for detecting mycotoxins in distiller's grains.

Looking Ahead

The FDA is working on developing several detection tools that are expected to be complete in 2008, including:

- Method for detecting melamine and cyanuric acid in various feeds and feed ingredients.
- Method for detecting antibiotics of concern in distiller's grains.
- Method for detecting prohibited proteins in animal feeds.

Response

3.1. Improve Immediate Response

Enhancing Traceability Capabilities for More Rapid and Precise Product Tracking

Accomplishments

- FDA is working to enhance traceability capabilities because the ability to quickly identify where a contaminated product came from and where it has been distributed is essential for a rapid response to a foodborne illness outbreak.
- FDA is currently engaging in dialogue with industry and other stakeholders on current traceability models and standards to identify best practices for traceability.

Looking Ahead

- FDA plans to develop guidance for traceability best practices to ensure fast and effective recalls.

Collaborating with States on Response Efforts

Accomplishments

FDA is working in conjunction with States to improve response to food safety threats.

- FDA has completed four Incident Command System training courses that have included State representatives.
- FDA has developed an additional Farm Investigation Course for federal, state, and international investigators. Training was held in February 2008.

Looking Ahead

- FDA issued a Request for Applications (RFA) for funding to establish state Rapid Response Teams to investigate foodborne illness outbreaks, perform trace backs of implicated foods and evaluate data from investigations to identify trends. Six states will be awarded cooperative agreements in FY 2008; an additional three States will be awarded funding in FY 2009.

3.2. Improve Risk Communications to the Public, Industry, and Other Stakeholders

Improving Recall Communication

Accomplishments

- FDA developed templates for recall communications and presented them to the FDA Risk Communications Advisory Committee for input in March 2008.

Looking Ahead

- FDA is currently exploring the use of multiple and targeted channels to quickly alert consumers of a threat to food safety.

CONCLUSION

Since the implementation of the Food Protection Plan in November 2007, FDA has been working to further build the foundations of a strong food safety system based on prevention, intervention and response. FDA has collaborated with stakeholders to improve preventive controls, signed agreements with foreign countries to ensure the safety of imported food, begun research on novel detection and mitigation methods, and provided training to improve response. FDA will continue to work collaboratively with foreign, federal, state, and local government partners, Congress, industry, consumer organizations, public health officials and others to ensure the full implementation of the Food Protection Plan and to ensure that America's food supply remains safe.

In: Food Protection and Safety
Editor: Lyman F. Thomasson
ISBN 978-1-60692-187-6

Chapter 7

FOOD PROTECTION PLAN: QUESTIONS AND ANSWERS*

U.S. Food and Drug Administration

Why did the FDA develop a food protection plan?

Societal and demographic changes, the globalization of the food supply, new threats and communication issues require a new approach to food protection. In May 2007, U.S. Department of Health and Human Services Secretary Michael O. Leavitt and Andrew C. von Eschenbach, M. D., Commissioner of Food and Drugs, charged the FDA with developing a comprehensive Food Protection Plan to address both unintentional and deliberate contamination of the nation's food supply.

Why is the FDA Issuing its Own Food Safety Plan? Isn't Food Safety Part of The Import Safety Action Plan?

The Food Protection Plan is integrated with the Administration's Import Safety Action Plan. The Food Protection Plan, which focuses on both domestic and imported food, is a comprehensive three-part plan that uses science and a risk-based approach of prevention, intervention and response to ensure the safety of

* Excerpted from U.S. FDA website http://www.fda.gov/oc/initiatves/advance/food/questions.html.

domestic as well as imported foods eaten by American consumers. This plan will help strengthen FDA's role as well as support the Agency's ongoing collaboration with other federal agencies that have a role in the safety of our nation's food supply, such as the Centers for Disease Control and Prevention and the USDA's Food Safety and Inspection Service.

What Strategy Does the Food Protection Plan Propose to Better Protect the U.S. Food Supply?

This approach is built on three core elements: Prevention, Intervention, and Response. Prevention from the start of the food cycle is the key to reducing food-borne contaminants from reaching consumers. Prevention, combined with intervention, will reduce the likelihood that contaminated products will reach the consumer. Prevention and Intervention are augmented with faster and focused Response once a problem occurs.

What New Legislation Does the Food Protection Plan Recommend?

The plan recommends that FDA request additional regulatory authorities to increase its capability to effectively address the three elements of the Food Protection Plan and meet the expectations of the American public.

Prevention

- Allow FDA to require preventive controls against intentional adulteration by terrorists or criminals at points of high vulnerability in the food chain
- Authorize FDA to issue additional preventive controls for high-risk foods
- Require food facilities to renew their FDA registrations every two years and allow FDA to modify the registration categories

Intervention

- Authorize FDA to accredit highly qualified third parties for voluntary food inspections ı Require new reinspection fee from facilities that fail to meet current Good Manufacturing Practices (cGMPS)
- Empower FDA to require electronic import certificates for shipments of designated high-risk products
- ı Require new Food and Animal Feed Export Certification Fee to improve the ability of U.S. firms to export their products
- ı Authorize FDA to refuse admission of imported food if FDA inspection access is delayed, limited or denied

Response

- Empower FDA to issue a Mandatory Recall if voluntary recalls are not effective
- ı Give FDA enhanced access to food records during emergencies

What Changes will the FDA Implement to Better Protect Our Food Supply?

The plan proposes numerous steps to meet the goals of the three basic components. The actions include, for example:

Prevention:

- Meet with states and consumer groups to solicit their input on implementing preventive approaches to protect the food supply
- ı Develop written food protection guidelines for industry to develop food protection plans for produce and other food products and implement other measures to promote corporate responsibility
- ı Analyze food import trend data and integrate it into a risk-based approach that focuses inspection resources on those imports that pose the greatest risk
- ı Improve FDA's presence overseas

Intervention

- Focus food and feed inspections and sampling based on risk
- Train FDA and state investigators on new, technically complex and specialized food manufacturing processes, as determined by a risk-based needs assessment, and modern inspection strategies
- Collaborate with foreign authorities to reduce potential risk of imported foods

Response

- Enhance data collection, incident reporting and emergency response capabilities
- Work with stakeholders to implement a more effective trace-back process, using technologies to rapidly and precisely track the origin and destination of contaminated foods, feed and ingredients
- Work with communications and media experts to design and conduct consumer communications and behavior response studies
- In a food-related emergency, implement this communications plan, including using all relevant media and technologies available, to reach consumers, retailers, industry, public health officials and other stakeholders

Are there More Food-Borne Illnesses Today thanthere were in the Past?

There is no solid evidence that the total number of food-borne illnesses have increased in recent years.

What are the Common Symptoms of Food-Borne Illnesses?

The symptoms of food-borne illness range from mild gastroenteritis to life-threatening neuralgic, hepatic and renal syndromes. These extend from short-term discomfort to hospitalization and even death.

How Soon will the FDA Start Implementing the Food Protection Plan?

Implementation of the Food Protection Plan will begin immediately, be phased in over time and integrated with the Administration's Import Safety Action Plan.

Why is FDA Requesting Authority to Require Mandatory Preventive Controls for Intentional Contamination Now?

Since the terrorist attacks of Sept. 11, 2001, FDA has solidified its initial work analyzing security in the food sector using a risk assessment tool known as CARVER+Shock. By requiring some practical food defense measures at specific points in the food supply chain, we can increase the security of the food supply. This process will be accompanied by significant opportunity for public comment and input and take into account the best available understanding of the uncertainties, risks, costs and benefits associated with alternative options.

Will FDA be Able to Prevent Food-Borne Illnesses With this Plan?

FDA will never be able to prevent 100% of food-borne illness. However, prevention is the keystone of this food protection plan and building prevention in up front is critical to better ensure safe food. This plan will further develop tools and science necessary to better understand where vulnerabilities are and the most effective ways to minimize them.

How can FDA Determine the Food Areas at Greatest Risk?

All risk-based approaches must be based on sciencc. The plan addresses the food product life cycle, from production to consumption. By analyzing data collected throughout the life cycle of domestic and imported food, FDA is better able to detect risks posed by food products as well as the key points where those risks can be reduced or eliminated through timely intervention.

FDA is developing a risk-based approach that accounts for variables such as:

- The likelihood that a certain food caused the illness

- The severity of the illness
- The point in the production process where contamination is most likely to occur
- The steps taken during production to reduce the likelihood of contamination.

When it Comes to Imported Foods, How Will the FDA Focus on 'Prevention?'

The concept is to "push the border out" and build safety in at the point of production and manufacture in other countries. FDA plans to work with the governments of other countries, industry and other partners to ensure that foods produced outside the U.S. provide the same level of protection as domestically-produced food.

How will the FDA Intervene to Eliminate Risks in the U.S. Food Supply?

Prevention is the key. To enforce preventive measures to drastically minimize risk, FDA will conduct targeted inspections and testing based on risk and using advanced science and technology of the 21st century. New detection technology is being designed for field use to allow for faster turn around of samples.

As Part of its Risk-Based Intervention, what Action is FDA Planning to Take Against Companies that Do not Implement Preventive Controls?

FDA could use its existing enforcement tools, such as untitled and warning letters, seizure, injunction and criminal prosecution for failure to implement any preventive controls that are mandatory.

How Does the FDA Receive Word of a Potential Problem?

There are a variety of ways the FDA is alerted to potential problems with products: consumer complaints, inspection data, test results, adverse event reports and reports of illness. We might also receive word of a problem from industry, one of our state or federal partners or from governments of other countries.

How Does FDA Plan to Improve Response Time to the Consumers when there is an Outbreak or Recall?

By improving FDA's ability to detect early signs of problems combined with improved data integration and the development of more response teams at a state level FDA will be able to respond faster. FDA will also look for better ways to communicate with consumers and retailers during food emergencies.

What is the FDA Doing to Improve the Communication Among other Federal and State Agencies to Ensure the Safety of U.S. Food Products?

The FDA communicates regularly with other federal agencies that are involved in safeguarding the U.S. food supply including USDA, DHS, and CDC. FDA also communicates on a regular basis with the states through quarterly 50-state conference calls and as needed as a result of emergencies, food-borne outbreaks or significant events through our Division of Federal–State Relations. Other federal agencies such as USDA, CDC, and DoD are invited to participate in these calls. FDA has regular interactions with CDC regarding ongoing food-borne illness situations and surveillance that is facilitated by having an on-site FDA liaison situated at CDC in Atlanta.

Communications such as recalls, FDA press releases, health alerts and significant state notices are broadcast through the SAFE system (State Advisory Fax-Email system) to over 2,500 fax/email addresses. Today, more than 10,000 state food contract inspections are entered into the eSAF (electronic State Access to FACTS) and FDA is working with the states to determine ways to enter and access another 40,000 inspections conducted by the states under their state laws and programs.

The use of other state data for risk management and enforcement to complement FDA data also is under consideration. In addition, FDA is piloting the Manufactured Food Regulatory Program

Standards, a set of best practices for state regulatory programs, with five states in FY08 with expansion planned to all state contract programs by 2011.

What are Some of the New Scientific Advances that will be Used to Improve Response Time in Handling Food-Borne Illnesses?

The development of new technology, along with the implementation of technology that is already in development, will allow inspectors to identify risk-based factors faster. These tools include realtime diagnostic instruments and methods that allow for rapid, on-site analysis of a particular sample. This type of technology could, in some cases, reduce analysis time from days to minutes.

What Role will Other Agencies Have in Protecting the Food Supply (USDA FSIS, CDC, NOAA, Etc.)?

While FDA has responsibility for all types of food with the exception of meat, poultry and egg products, which are under the jurisdiction of the Department of Agriculture (USDA), protecting the food supply has always been an integrated approach across multiple federal agencies. The Food Protection Plan addresses FDA's role but will require the continued integration and interaction with multiple federal, state and local regulatory authorities.

Is FDA Asking for Authority to Require that a Stamp of Approval be Placed on All Imported Products?

No, FDA is seeking authority to require certificates of compliance with FDA standards for designated products from specific countries based on agreements with those countries. Products will be designated for such a certificate requirement based on risk.

How will the Proposed New Authorities Extend FDA's Presence on Farms?

FDA currently has authorities applicable to food on farms, such as its authorities to perform inspections and seize food. With the exception of specific situations related to preventing deliberate contamination of food (i.e., milk) FDA is not requesting new authority that will extend FDA's presence on farms.

In: Food Protection and Safety
Editor: Lyman F. Thomasson
ISBN 978-1-60692-187-6

Chapter 8

SALMONELLOSIS OUTBREAK FROM CERTAIN TOMATOES. QUESTIONS AND ANSWERS FOR CONSUMERS AND INDUSTRY*

U.S. Food and Drug Administration

CONSUMER INFORMATION AND ADVICE

- What states have confirmed cases of Salmonella illness linked to this outbreak?
- How many people have become ill in this outbreak?
- What kind of illness does Salmonella cause?
- What is Salmonella?
- What kinds of raw tomatoes should be avoided during this outbreak?
- What kinds of raw tomatoes should consumers continue to buy during this outbreak?
- What states or countries have been ruled out as the source of this outbreak?
- How can consumers tell where a tomato was grown, harvested, or packed?

* Excerpted from U.S. FDA Report dated June 9, 2008; updated June 21, 2008.

- Will washing the tomatoes implicated in this outbreak make them safe to eat?
- Will peeling the tomatoes implicated in this outbreak make them safe to eat?
- Can cooking tomatoes eliminate Salmonella?
- Are canned tomatoes and processed foods containing tomatoes safe for consumers during this outbreak?
- Are tomatoes from farmers' markets included in this outbreak?
- During this outbreak, is it safer to eat locally grown tomatoes?
- What steps are being taken to prevent future illness from this outbreak?
- How should consumers handle raw tomatoes NOT implicated in this outbreak?
- Where can consumers find out more about how to keep from becoming sick from tomatoes?

Advice for Food-Service Providers, Restaurateurs, and Retailers

- What is FDA's advice to retailers, restaurateurs, and food service providers?
- What kinds of tomatoes may be offered to customers?
- What safe-handling practices should be followed for tomatoes?
- Should food-service providers, restaurateurs, and retailers cook the tomatoes associated with the outbreak to eliminate Salmonella?
- Should food-service providers, restaurateurs, and retailers try to remove Salmonella by washing the tomatoes associated with the outbreak?

About Outbreaks

- What is an outbreak?
- When did the illnesses associated with the current outbreak start?
- How is the cause or source of a Salmonella outbreak determined?
- What is the FDA doing to identify the source of this outbreak?

- Why is it taking FDA so long to determine the source of this Salmonella outbreak?
- From farm to table, where in the process are tomatoes most likely to become contaminated? What are the most likely sources of contamination?
- Have there been other outbreaks from contaminated tomatoes, before this one?

Government Activities Related to Produce Safety

- What steps has the FDA taken to reduce the potential for Salmonella outbreaks from tomatoes?
- Does the FDA sample and test domestic and foreign tomatoes?
- On the list of areas not associated with this outbreak, there is a special notation for certain counties in Florida. How is the FDA guaranteeing consumers that tomatoes are from those areas?
- Has the FDA conducted outreach/education activities regarding tomato safety?
- What is the FDA's Food Protection Plan?

Consumer Information and Advice

What States Have Confirmed Cases of Salmonella Illness Linked to this Outbreak?

As of June 17, 2008, confirmed cases had been reported in Arizona, Arkansas, California, Colorado,Connecticut, Florida, Georgia, Idaho, Illinois, Indiana, Kansas, Kentucky, Maryland, Michigan, Missouri, New Mexico, North Carolina, New York, Ohio, Oklahoma, Oregon, Tennessee, Texas, Utah, Virginia, Vermont, Washington, and Wisconsin, and in the District of Columbia.

How Many People Have Become Ill in this Outbreak?

The current numbers of illnesses and hospitalizations that have been reported are continually updated elsewhere on the FDA web site.

What Kind of Illness Does Salmonella Cause?

People who have eaten food contaminated with Salmonella often have fever, diarrhea (which may be bloody), nausea, vomiting, and abdominal pain. The bacterium can enter the bloodstream and cause more severe illness, although this rarely happens. Infection with Salmonella also may be more serious or fatal in young children, frail or elderly people, and people with weakened immune systems.

What is Salmonella?

Salmonella is a type of bacterium. The type of Salmonella causing illness in this outbreak, Salmonella Saintpaul, is relatively uncommon. Fruits and vegetables that come into contact with Salmonella may become contaminated with it, causing illness if eaten. Salmonella lives in the intestinal tracts of some animals, and can live in soil and water for months. Once Salmonella has contaminated something, it can be spread from surface to surface. A tomato contaminated with Salmonella can spread the bacterium to the hands of a person who cuts the tomato and to the cutting board on which the tomato is sliced, for example.

What Kinds of Raw Tomatoes Should be Avoided During this Outbreak?

At this time, consumers should avoid eating or handling raw red plum, raw red Roma, and raw red round tomatoes, and foods containing them, unless they are from regions that have been ruled out as the sources of the contamination. If consumers already have these kinds of tomatoes in their homes and are unsure where they were grown or harvested, they are encouraged to contact the store where they bought the tomatoes. (Check the list on the FDA website often, because regions continue to be added as they are ruled out as the sources of the outbreak).

What Kinds of Raw Tomatoes should Consumers Continue to Buy During this Outbreak?

Consumers may continue to buy any type of tomato from sources that have NOT been linked to the outbreak and foods that contain them. In addition, raw cherry tomatoes, grape tomatoes, and tomatoes with the vine still attached have not been linked to the outbreak in any region, and consumers may also continue to buy them. Tomatoes grown at home also are not linked to this outbreak.

What States or Countries Have been Ruled out as the Source of This Outbreak?

A current list of states, territories, and countries that have been ruled out as the source of this outbreak is available elsewhere on the FDA website.

The U.S. states in which people have become ill from tomatoes are not necessarily the areas where the tomatoes were grown. The tomatoes might have been shipped to these states from elsewhere. Check this list on the FDA website often, as more regions that have been ruled out as the sources of the outbreak are being added.

How Can Consumers Tell where a Tomato Was Grown, Harvested, or Packed?

Consumers can ask retailers (for example, store and restaurant personnel) where their raw red plum tomatoes, raw red Roma tomatoes, and raw red round tomatoes were produced. If consumers have any doubts about where these types of tomatoes were grown, harvested, or packed, they should discard them. Because grape tomatoes, cherry tomatoes, and tomatoes sold with the vine still attached are not implicated in this outbreak, regardless of where they are from, it's not necessary to ask where they were grown, harvested, or packed.

Will Washing the Tomatoes Identified in this Outbreak Make them Safe to Eat?

Consumers are advised not to try to wash raw red plum, red Roma, or raw red round tomatoes that are implicated in the outbreak. Consumers should throw these tomatoes out. Once produce is contaminated with pathogens such as Salmonella, the contamination is difficult to remove.

Will Peeling the Tomatoes Implicated in this Outbreak Make them Safe to Eat?

Consumers are advised not to try to peel raw red plum, red Roma, or raw red round tomatoes that have not been ruled out as the source of the outbreak. Consumers should throw these tomatoes out. Peeling a tomato that is contaminated on the outside would be likely to spread the contamination to the inside. Also, if the contamination is already on the inside, peeling will not remove it.

Can Cooking Tomatoes Eliminate Salmonella?

Consumers should not attempt to cook the tomatoes involved in this outbreak in an effort to kill Salmonella. Handling tomatoes contaminated with Salmonella can spread the bacterium to anything the handler touches, including hands, kitchen utensils, cutting boards, sinks, and other foods. Cooking tomatoes in the home will not ensure that Salmonella is eliminated.

Are Canned Tomatoes and Processed Foods Containing Tomatoes Safe for Consumers During this Outbreak?

Consumers may continue to buy and eat canned or bottled (that is, processed) red plum, red Roma, and red round tomatoes and canned or bottled foods containing these or any other types of tomatoes if they were processed by a commercial food-processing facility. A few examples are the canned tomatoes and canned or bottled tomato juice and spaghetti sauce found in grocery stores.

Are Tomatoes from Farmers' Markets Included in this Outbreak?

Farmers' markets get their tomatoes from a variety of sources that are not necessarily limited to local farms. These other sources may include the same ones that provided the tomatoes implicated in the Salmonella outbreak. Consumers should ask retailers at farmers' markets what the sources of their tomatoes are, and frequently check FDA updates for states, territories, and countries not associated with the outbreak.

During this Outbreak, is it Safer to Eat Locally Grown Tomatoes?

Consumers should confirm with their retailers the sources of tomatoes advertised as "locally grown." Consumers also are reminded that cherry tomatoes, grape tomatoes, and tomatoes with the vine still attached are not implicated in this outbreak, regardless of where they were grown, harvested, or packed. Tomatoes consumers grow at home also are not implicated in the outbreak.

What Steps are Being Taken to Prevent Future Illness From this Outbreak?

The FDA, the Centers for Disease Control and Prevention (CDC), the Indian Health Service, and state health department officials are collaborating closely to ensure that the outbreak is contained and that consumers and retailers are made aware of the contamination. The food industry is cooperating in the investigation and is assisting government officials in their efforts to identify distribution patterns and to find the source of the contamination, to ensure that additional contaminated tomatoes do not reach consumers.

How should Consumers Handle Raw Tomatoes NOT Associated with the Outbreak?

- Again, the tomatoes associated with the outbreak should be thrown out. For tomatoes not associated with the outbreak, consumers should follow the usual recommendations:
- Don't buy or eat tomatoes that look damaged; for example, if the skin of a tomato is broken or the tomato is spoiled, the tomato should be thrown out.
- Stored tomatoes should not come in contact with raw meat, poultry, or eggs.
- Wash hands with soap and warm water before handling tomatoes.

- Wash each tomato thoroughly under running water. Don't wash tomatoes in a tub or sink filled with water.
- When finished washing a tomato, cut out the scar where the stem was, and throw it away. o Never cut a fresh tomato until it has been thoroughly washed.
- Cut the tomato on a clean cutting board, using clean utensils. Don't let the tomato come in contact with other raw foods, including raw meat, poultry, and eggs, or the surfaces they have touched.
- Wash cutting boards and utensils in between each different type of food that is cut.
- Refrigerate fresh, cut tomatoes (or products made from them, such as salsa) at 41° F or less.
- (Note: Refrigeration will not kill Salmonella that is already present on tomatoes.)
- Wash hands with soap and warm water after preparing the tomatoes.

The FDA does not recommend using any kinds of detergents to wash fresh produce, because it is not yet known if their residues are harmful to humans.

Where can Consumers Find out More about How to Keep From Becoming Sick from Tomatoes?

More information about safe handling of fresh produce is available on the FDA website. Advice for Food Service Providers, Restaurateurs, and Retailers

What is the FDA's Advice to Retailers, Restaurateurs, and Food Service Providers?

Food-service providers, restaurateurs, and retailers should discard raw red plum tomatoes, raw red Roma tomatoes, and raw red round tomatoes, unless they were grown, harvested, and packed in one of the regions not associated with the outbreak. Check the list often for updates.

What Kinds of Tomatoes May be Offered to Customers?

Food service providers, restaurateurs, and retailers may continue to offer ANY raw tomatoes from areas not implicated in the outbreak. Regardless of where they are from, raw cherry tomatoes, raw grape tomatoes, and raw tomatoes sold with the vine still attached also may be offered, as they have not been associated with this outbreak. Foods containing any type of tomato, from any area, that have been bottled or canned in industrial food-processing facilities may be offered; for example, the type of canned tomatoes, tomato sauce, or vegetable juice found in grocery stores.

What Safe-Handling Practicesshould be Followed for Tomatoes?

Tomatoes associated with this outbreak should be discarded. For the types of raw tomatoes not associated with the outbreak, retailers, restaurateurs, and food service providers should follow the safe-handling practices listed above for consumers. Additional details about refrigeration practices are shown below. (Note that refrigeration will not eliminate Salmonella that is already present on a tomato.)

- Fresh, cut tomatoes should be refrigerated at 41° F or less.
- If fresh, cut tomatoes are added to another food, such as salsa, the food should be refrigerated at 41° F or less.
- If fresh, cut tomatoes are placed on a salad, it should be refrigerated at 41° F or less. o Fresh, cut tomatoes on sandwiches should be refrigerated at 41° F or less.

Should Food-Service Providers, Restaurateurs, and Retailers Cook the Tomatoes Associated with the Outbreak to Eliminate Salmonella?

No. Food-service providers, restaurateurs, and retailers should not attempt to cook the tomatoes involved in this outbreak in an effort to kill Salmonella, because the potential for cross-contamination is high. The FDA recommends discarding the tomatoes implicated in the outbreak, unless the tomatoes were produced in one of the regions that have been ruled out as sources of the contamination.

Should Food-Service Providers, Restaurateurs, and Retailers try to Remove Salmonella by Washing the Tomatoes Associated with the Salmonella Outbreak?

No. Food-service providers, restaurateurs, and retailers should not attempt to remove Salmonella by washing. Salmonella has physical properties that make it difficult to remove once it is present on a tomato. Conventional cleaning methods are not likely to completely eliminate it. In addition, the potential for cross-contamination is high. The FDA recommends discarding the tomatoes implicated in the outbreak, unless the tomatoes were produced in one of the regions that have been ruled out as sources of the contamination.

ABOUT OUTBREAKS

What is an Outbreak?

An outbreak is defined by the CDC as two or more cases of the same disease that share a common exposure.

When did the Illnesses Associated with the Current Outbreak Start?

The illnesses began in mid-April and continue to be reported.

How is the Cause or Source of aSalmonella Outbreak Determined?

Once an outbreak is detected and the states and the CDC have determined that two or more cases of the same disease share a common food exposure, and the food is identified, the FDA conducts a "trace-back" investigation to determine the source of the contaminated food. The product is tracked from the point of purchase or service through each point in the distribution chain to find the source of the contamination.

At each point in the distribution chain, an environmental investigation is performed to determine whether the contamination may have occurred at that point and, if so, how it occurred. When outbreak illnesses occur across multiple states, the contamination often occurred at, or near, the original source of the product, such as the growing or packing area. In addition to helping to contain current outbreaks, information gained from trace-back and other investigations can help scientists develop measures to prevent future occurrences.

What is the FDA Doing to Identify the Source of this Outbreak?

The FDA is conducting a trace-back investigation. Epidemiological information about the disease serotype (Salmonella Saintpaul serotype) is being examined, disease patterns are being linked, and seasonal distribution patterns in the marketplace are being analyzed to rule out sources.

The federal (principally CDC and the FDA) and state governments continue to work together to analyze samples from ill persons and product samples of tomatoes. The strain of Salmonella from ill persons is being "fingerprinted" at public health laboratories around the country, as part of PulseNet (the network of public health laboratories that sub-type bacteria). All Salmonella strains associated with this outbreak have the same genetic "fingerprint" (DNA pattern).

Why is it Taking FDA So Long to Determine the Source of this Salmonella Outbreak?

Investigators must track the pathways that the tomatoes associated with illness followed, from multiple consumers who ate them to the multiple retailers or restaurants that sold them; from there to multiple points of supply and distribution; to where the tomatoes were packed, and to where they were harvested and grown. At the points where the tomatoes were sold or prepared, investigators try to determine identifying information, such as packaging, labeling, and lot numbers; when the tomatoes were purchased or prepared, and what the receiving, stock-rotation, inventory, handling, and shipping procedures were. They collect records about suppliers and shipments to retailers or restaurants for the period of the tomatoes' shelf lives. Investigators then chart and analyze distribution data, accomplished by tracing lot numbers - if they are available - or by using a shipment-delivery timeline to determine if the tomatoes were useable and "sellable" during the period of infection.

Distributor interview, data collection, and analysis are repeated for multiple levels of distribution until the source of the tomatoes is identified.

Among the complications that arise in this process is that lot numbers and other information identifying the tomatoes' growers might not be included on receipts and shipping records. In some cases, investigators have to rely on reviewing records and interviewing the personnel who handle such matters, which increases the time and resources needed to trace implicated tomatoes back to their sources. Another complication that delays the investigation is that often there is no package, no product code, no "sell by" date, and no marking on the tomato at the retail level.

For more information about this process, visit the Guide to Traceback of Fresh Fruits and Vegetables Implicated in Epidemiological Investigations that FDA has posted on its web site.

From Farm to Table, where in the Process are Tomatoes Most Likely to Become Contaminated? What are the Most Likely Sources of Contamination?

Fresh produce, including tomatoes, can become contaminated at any point along the supply chain, from the field or greenhouse where it is grown to distribution points to food preparation in restaurants and homes.

The FDA's 1998 Guide to Minimize Microbial Contamination of Fresh Fruits and Vegetables (also referred to as the Good Agricultural Practices (GAPs) guide) describes potential sources of microbial contamination in the field and packing house environments and makes recommendations for how to reduce or minimize opportunities for contamination.

According to the GAPs guide, areas that should be considered to minimize the potential for the microbial contamination of produce include agricultural water (e.g., for irrigation or crop protection sprays); wild and domestic animals; worker health and hygiene; the production environment (use of manure, previous land use, and use of adjacent land); post-harvest water quality (water used to wash or cool produce) and sanitation of facilities and equipment.

Have there been other Outbreaks From Contaminated Tomatoes, before this One?

Since 1990, at least 13 large, multi-state foodborne outbreaks and some small local outbreaks have been associated with different varieties of tomatoes. From 1998 to 2006, outbreaks reported to the FDA that were associated with tomatoes made up 17 percent of produce-related outbreaks. Salmonella has been the pathogen of concern most often associated with outbreaks from tomatoes.

GOVERNMENT ACTIVITIES RELATED TO PRODUCE SAFETY

What Steps has the FDA Taken to Reduce the Potential for Salmonella Outbreaks from Tomatoes?

On June 12, 2007, the FDA announced a Tomato Safety Initiative, a multi-year effort focusing on the East Coast. The Initiative is a collaborative effort between the FDA and the state health and agriculture departments in Virginia and Florida, in cooperation with several universities and the produce industry. This initiative is part of an ongoing, preventive, risk-based strategy.

The Tomato Safety Initiative includes identifying practices or conditions that potentially lead to contamination of tomatoes, and what steps producers are taking to address these issues. Information from the Initiative will allow the FDA to continue to improve its guidance and policy on tomato safety. The Initiative also is evaluating the need for additional produce safety research, education, and outreach. The Initiative supports an important goal in the 2004 FDA Produce Safety Action Plan – minimizing the incidence of foodborne illness associated with the consumption of fresh produce – and the prevention activities described in the FDA's Food Protection Plan.

Does FDA Sample and Test Domestic and Foreign Tomatoes?

The FDA routinely collects random samples of tomatoes of all varieties, domestic and imported, from various growers, packers and shippers. The samples are sent to a FDA laboratory, to be analyzed for a variety of bacteria, including Salmonella.

On the List of Areas not Associated with this Outbreak, there is a Special Notation for Certain Counties in Florida. How is the FDA Guaranteeing Consumers that Tomatoes are from fhose Areas?

The FDA evaluated and approved a proposal from the state of Florida that tomatoes grown and harvested in those counties could be offered for sale only if there was a certificate attached to each shipment that would follow it throughout its distribution. These certificates provide verification that the product coming from these counties was not harvested and shipped at the time of the onset of this outbreak. A unique numbering system is attached to these certificates, which must be issued by the state of Florida's Department of Agriculture.

Has the FDA Conducted Outreach/Education Activities Regarding Tomato Safety?

The FDA has issued a press release to notify the public of the current Salmonella outbreak in tomatoes; the press release is updated as information is obtained and evaluated. In addition, the FDA has posted consumer and industry (retailer) warnings and advice related to the current Salmonella outbreak on its website.

The FDA web site also includes a consumers' page about safe handling of fresh produce, including tomatoes. In 2006, the FDA issued a publication called Program Information Manual: Retail Food Protection — Storage and Handling of Tomatoes for members of the retail industry. Safe-handling guidelines for the tomato-supply industry are nearing completion.

What is the FDA's Food Protection Plan?

The FDA has developed a comprehensive Food Protection Plan to address the changes in food sources, production, and consumption we face in today's world. Building and improving on an already sound food- safety capability, the new plan is a strategy for protecting the nation's food supply. The plan approaches protection of the nation's food supply on three levels: prevention, intervention, and response. This new strategy will help ensure that Americans continue to benefit from one of the safest food supplies in the world.

In: Food Protection and Safety
Editor: Lyman F. Thomasson

ISBN 978-1-60692-187-6

Chapter 9

TOMATO SAFETY INITIATIVE*

U.S. Food and Drug Administration

The Food and Drug Administration (FDA) will begin a Tomato Safety Initiative in the Summer of 2007. The Initiative is a collaborative effort between FDA and the state health and agriculture departments in Virginia and Florida, in cooperation with several universities and members of the produce industry.

FDA developed the Tomato Safety Initiative in response to recurring Salmonella outbreaks associated with fresh and fresh-cut tomatoes. The Initiative is part of a risk-based strategy to reduce foodborne illness by focusing food safety efforts on specific products, practices, and growing areas that have been found to be problematic in the past. The Tomato Safety Initiative is modeled after the Leafy Greens Safety Initiative that was initiated in August 2006, in collaboration with the State of California's Department of Health Services and Department of Food and Agriculture. This new initiative is fully consistent with the 2004 FDA Produce Safety Action Plan goal of minimizing the incidence of foodborne illness associated with the consumption of fresh produce.

Most tomato-associated outbreaks over the past ten years have been traced to product originating from the Eastern shore of Virginia and from Florida; however outbreaks have also been traced to Georgia, South Carolina, Ohio, and California. Accordingly, FDA, in cooperation with Virginia's Department of Agriculture and Consumer Services, will begin the Initiative in July of this year, by visiting

* Excerpted from U.S. FDA Report dated June 12, 2007.

Virginia based tomato farms and packing facilities to assess their food safety practices and to what degree they implement Good Agricultural Practices (GAPs) and Good Manufacturing Practices (GMPs). Assessment of a variety of environmental factors including irrigation water, wells, procedures for mixing chemicals, drought and flooding events, and animal proximity to growing fields will also be conducted during the farm and packing facility visits. Later in the year, a similar effort will be conducted in Florida coinciding with the production and harvesting season there. Other components of the Initiative include continuing outreach with the industry at all points in the supply chain, facilitating and promoting research on tomato safety, and working with federal, state and local public health officials in disease detection and outbreak response.

By identifying practices or conditions that potentially lead to product contamination, FDA can further improve guidance and policy intended to minimize future outbreaks as well as ascertain future produce safety research, education, and outreach needs. The findings of the Initiative will be publicly shared upon completion of the effort, allowing the states and industry members to maximize their food safety efforts as well.

In: Food Protection and Safety
Editor: Lyman F. Thomasson
ISBN 978-1-60692-187-6

Chapter 10

FDA Implementing Initiative to Reduce Tomato-Related Foodborne Illnesses*

U.S. Food and Drug Administration

The U.S. Food and Drug Administration (FDA) will begin a multi-year Tomato Safety Initiative to reduce the incidence of tomato-related foodborne illness in the United States.

"Produce is an important part of a healthy diet and FDA wants to improve its safety by better understanding the causes of foodborne illness and by promoting more effective methods of safe food production, delivery, and preparation," said Robert Brackett, Ph.D., director of FDA's Center for Food Safety and Applied Nutrition. "This initiative is part of a strategy to reduce foodborne illness by focusing food safety assessments on specific products, practices, and growing areas that have been found to be problematic in the past."

The initiative, part of FDA's Produce Safety Action Plan, is a collaborative effort between FDA and state health and agriculture departments in Florida and Virginia. Several universities and members of the produce industry also are part of the effort. It will begin during this year's growing season for Virginia in the summer and for Florida in the fall.

During the past decade, the consumption of fresh and fresh-cut tomatoes has been linked to 12 different outbreaks of foodborne illness in the United States. Those outbreaks include 1,840 confirmed cases of illness. The majority of these

---

* Excerpted from U.S. FDA Report dated March 31, 2008.

outbreaks have been traced to products from Florida and the eastern shore of Virginia; however, tomato-associated outbreaks also have been traced to tomatoes from California, Georgia, Ohio, and South Carolina. The effort will include identifying practices or conditions that potentially lead to product contamination, which will allow FDA to continue to improve its guidance and policy on tomato safety. The initiative will evaluate the need for additional produce safety research, education, and outreach.

Other components of the initiative include:

- continuing outreach with the industry at all points in the supply chain,
- facilitating and promoting research on tomato safety,
- communicating early and often in the event of an outbreak, and
- continuing to build and strengthen collaborative relationships with federal, state and local public health officials in disease prevention, detection, and outbreak response.

FDA investigators in coordination with their respective state counterparts will visit tomato farms and packing facilities in Florida and Virginia to assess food safety practices and use of Good Agricultural Practices (GAPs) and Good Manufacturing Practices (GMPs). During their visits, officials will also evaluate a variety of environmental factors including irrigation water, wells, procedures for mixing chemicals, drought and flooding events, and animal proximity to growing fields.

The Tomato Safety Initiative complements the Leafy Greens Initiative launched in 2006.

In: Food Protection and Safety
Editor: Lyman F. Thomasson ISBN 978-1-60692-187-6

Chapter 11

TOMATO SAFETY INITIATIVE CONTINUES IN 2008

U.S. Food and Drug Administration

On March 31, 2008, FDA, in cooperation with Florida's Department of Agriculture and Consumer Services, Florida's Department of Health, the University of Florida, and CDC, began visiting Florida based tomato farms and packing facilities to assess conditions and practices and to what degree they implement Good Agricultural Practices and Good Manufacturing Practices (GAPs/GMPs). This is a continuation of the Tomato Safety Initiative launched in the summer of 2007. Assessments of Florida farms and packing facilities were originally planned for the fall of 2007, but were rescheduled to better coincide with the peak production season.

INDEX

A

B

C

D

G

H

I

N

O

P

Q

R

S

T

U

V

W